はじめに

　本章は、OpenStack Foundation が公開している公式ドキュメント「OpenStack Installation Guide for Ubuntu 14.04」の内容から、「Block Storage Service」までの構築手順をベースに加筆したものです。

　第 I 部 OpenStack 構築編では、OpenStack を Ubuntu Server 14.04 ベースで構築する手順を解説しています。Canonical 社が提供する Cloud Archive リポジトリーを使って、OpenStack の最新版 Mitaka を導入しましょう。

　第 II 部 監視環境構築編では、構築した OpenStack 環境を Zabbix と Hatohol で監視しましょう。Zabbix は Zabbix SIA 社が開発・提供・サポートする、オープンソースの監視ソリューションです。Hatohol は Project Hatohol が開発・提供する、システム監視やジョブ管理やインシデント管理、ログ管理など、様々な運用管理ツールのハブとなるツールです。Hatohol は Zabbix や Nagios、OpenStack Ceilometer に対応しており、これらのツールから情報を収集して性能情報、障害情報、ログなどを一括管理することができます。Hatohol のエンタープライズサポートはミラクル・リナックス株式会社が提供しています。本編では OpenStack 環境を監視するために Zabbix と Hatohol を構築するまでの流れを説明します。

OpenStack
構築手順書 [Mitaka版]

日本仮想化技術株式会社 = 著

購入者限定
FAQフォーラム
参加特典付

OSSクラウド環境基盤の大本命！
<Ubuntu Server 14.04 ベース>
Keystone / Glance / Nova / Neutron / Cinder

インプレス

- 本書は、日本仮想化技術が運営するEnterpriseCloud.jpで提供している「OpenStack構築手順書」をオンデマンド書籍として再編集したものです。
- 本書の内容は、執筆時点までの情報を基に執筆されています。紹介したWebサイトやアプリケーション、サービスは変更される可能性があります。
- 本書の内容によって生じる、直接または間接被害について、著者ならびに弊社では、一切の責任を負いかねます。
- 本書中の会社名、製品名、サービス名などは、一般に各社の登録商標、または商標です。なお、本書では©、®、TMは明記していません。

目　次

はじめに iii

第 I 部　OpenStack 構築編　　1

第 1 章　構築する環境について 3
1.1　環境構築に使用する OS 3
1.2　作成するサーバー（ノード）... 4
1.3　ネットワークセグメントの設定 4
1.4　各ノードのネットワーク設定 5
1.5　Ubuntu Server のインストール 5
1.6　Ubuntu Server へのログインと root 権限 7
1.7　設定ファイル等の記述について 7

第 2 章　OpenStack インストール前の設定 9
2.1　ネットワークデバイスの設定 9
2.2　ホスト名と静的な名前解決の設定 10
2.3　リポジトリーの設定とパッケージの更新 11
2.4　OpenStack クライアントと MariaDB クライアントのインストール 12
2.5　時刻同期サーバーのインストールと設定 12
2.6　MariaDB のインストール 13
2.7　RabbitMQ のインストール 14

2.8	環境変数設定ファイルの作成	16
2.9	memcached のインストールと設定	16

第 3 章　Keystone のインストールと設定（コントローラーノード）　19

3.1	データベースを作成	19
3.2	データベースの確認	19
3.3	admin_token の決定	20
3.4	パッケージのインストール	20
3.5	Keystone の設定変更	20
3.6	データベースに展開	21
3.7	Fernet キーの初期化	21
3.8	Apache Web サーバーの設定	21
3.9	サービスの再起動と不要 DB の削除	22
3.10	サービスと API エンドポイントの作成	23
3.11	プロジェクトとユーザー、ロールの作成	24
3.12	Keystone の動作確認	26

第 4 章　Glance のインストールと設定　29

4.1	データベースを作成	29
4.2	データベースの確認	29
4.3	ユーザー、サービス、API エンドポイントの作成	30
4.4	Glance のインストール	31
4.5	Glance の設定変更	31
4.6	データベースに展開	32
4.7	Glance サービスの再起動	33
4.8	ログの確認と使用しないデータベースファイルの削除	33
4.9	イメージの取得と登録	33

第 5 章　Nova のインストールと設定（コントローラーノード）　35

5.1	データベースを作成	35
5.2	データベースの確認	35
5.3	ユーザーとサービス、API エンドポイントの作成	36

5.4	パッケージのインストール	37
5.5	Nova の設定変更	37
5.6	データベースに展開	39
5.7	Nova サービスの再起動	39
5.8	不要なデータベースファイルの削除	39
5.9	Glance との通信確認	39

第 6 章　Nova Compute のインストールと設定（コンピュートノード）　41

6.1	パッケージのインストール	41
6.2	Nova の設定を変更	41
6.3	Nova コンピュートサービスの再起動	43
6.4	コントローラーノードとの疎通確認	43

第 7 章　Neutron のインストール・設定（コントローラーノード）　45

7.1	データベースを作成	45
7.2	データベースの確認	45
7.3	neutron ユーザーとサービス、API エンドポイントの作成	46
7.4	パッケージのインストール	47
7.5	Neutron コンポーネントの設定	47
7.6	Nova の設定を変更	52
7.7	データベースに展開	53
7.8	コントローラーノードの Neutron と関連サービスの再起動	53
7.9	ログの確認	54
7.10	使用しないデータベースファイルを削除	54

第 8 章　Neutron のインストール・設定（コンピュートノード）　55

8.1	パッケージのインストール	55
8.2	設定の変更	55
8.3	コンピュートノードのネットワーク設定	57
8.4	コンピュートノードの Neutron と関連サービスを再起動	57
8.5	ログの確認	57
8.6	Neutron サービスの動作を確認	58

第 9 章　仮想ネットワーク設定（コントローラーノード） 59
- 9.1　パブリックネットワークの設定 60
- 9.2　インスタンス用ネットワークの設定 61
- 9.3　仮想ネットワークルーターの設定 63
- 9.4　ネットワークの確認 64
- 9.5　インスタンスの起動を確認 65

第 10 章　Cinder のインストール（コントローラーノード） 69
- 10.1　データベースを作成 69
- 10.2　データベースの確認 69
- 10.3　Cinder サービスなどの作成 70
- 10.4　パッケージのインストール 71
- 10.5　Cinder の設定を変更 71
- 10.6　データベースに展開 72
- 10.7　使用しないデータベースファイルを削除 73
- 10.8　nova-api の設定変更 73
- 10.9　サービスの再起動 73
- 10.10　イメージ格納用ボリュームの作成 73

第 11 章　Dashboard のインストールと確認（コントローラーノード） 77
- 11.1　パッケージのインストール 77
- 11.2　Dashboard の設定を変更 77
- 11.3　Dashboard にアクセス 78
- 11.4　セキュリティグループの設定 79
- 11.5　キーペアの作成 79
- 11.6　インスタンスの起動 79
- 11.7　Floating IP の設定 80
- 11.8　インスタンスへのアクセス 80
- 11.9　Ubuntu テーマの削除 81

第 II 部　監視環境 構築編　　83

第 12 章　Zabbix のインストール　　85
12.1　パッケージのインストール　　85
12.2　Zabbix 用データベースの作成　　86
12.3　Zabbix サーバーの設定および起動　　87
12.4　Zabbix frontend の設定および起動　　87
12.5　Zabbix frontend の日本語化設定　　89

第 13 章　Hatohol のインストール　　91
13.1　インストール　　91
13.2　MariaDB サーバーの設定　　92
13.3　セキュリティ設定の変更　　94
13.4　Hatohol Arm Plugin Interface 2 (HAPI2) の設定　　95
13.5　Hatohol による監視情報の閲覧　　96
13.6　Hatohol に Zabbix サーバーを登録　　97
13.7　Hatohol で Zabbix サーバーの監視　　97
13.8　Hatohol でその他のホストの監視　　98

付録 A　FAQ フォーラム参加特典について　　103

第Ⅰ部

OpenStack 構築編

第1章　構築する環境について

1.1　環境構築に使用するOS

　本書はCanonicalのUbuntu ServerとCloud Archiveリポジトリーのパッケージを使って、OpenStack Mitakaを構築する手順を解説したものです。

　Ubuntu Serverでは新しいハードウェアのサポートを積極的に行うディストリビューションです。そのため、Linux Kernelのバージョンを Trustyの場合は14.04.2以降のLTSのポイントリリースごとに、スタンダート版のUbuntuと同様のバージョンに置き換えてリリースしています。

- https://wiki.ubuntu.com/Kernel/LTSEnablementStack

　一般的な利用では特に問題ありませんが、OpenStackとSDNのソリューションを連携した環境を作る場合などに、Linux KernelやOSのバージョンを考慮しなくてはならない場合があります。また、Trustyの場合はLinux Kernel v3.13以外のバージョンのサポート期間はTrustyのサポート期間と同様ではありません。期限が切れた後はLinux Kernel v3.13にダウングレードするか、新しいLinux Kernelをインストールすることができるメタパッケージを手動で導入する必要がありますので注意してください。

　本書ではサポート期間が長く、Trustyの初期リリースの標準カーネルであるLinux Kernel v3.13を使うために、以下のURLよりイメージをダウンロードしたUbuntu Server 14.04.1 LTS(以下Ubuntu Server)のイメージを使ってインストールします。インストール後 apt-get dist-upgradeを行って最新のアップデートを適用した状態にしてください。Trustyではこのコ

マンドを実行してもカーネルのバージョンがアップグレードされることはありません。

本書は 3.13.0-86 以降のバージョンのカーネルで動作する Ubuntu 14.04.4 を想定しています。

- http://old-releases.ubuntu.com/releases/14.04.1/ubuntu-14.04.1-server-amd64.iso

> 筆者注：もしここで想定する Ubuntu やカーネルバージョン以外で何らかの事象が発生した場合も、以下までご報告ください。動作可否情報を Github で共有できればと思います。
> https://github.com/virtualtech/openstack-mitaka-docs/issues

1.2 作成するサーバー（ノード）

本書は OpenStack 環境を Controller,Compute の 2 台のサーバー上に構築することを想定しています。

コントローラー	コンピュート
RabbitMQ	Linux KVM
NTP	Nova Compute
MariaDB	Linux Bridge Agent
Keystone	
Glance	
Nova	
Neutron Server	
Linux Bridge Agent	
L3 Agent	
DHCP Agent	
Metadata Agent	
Cinder	

1.3 ネットワークセグメントの設定

IP アドレスは以下の構成で構築されている前提で解説します。

インターフェース	eth0
ネットワーク	10.0.0.0/24
ゲートウェイ	10.0.0.1
ネームサーバー	10.0.0.1

1.4 各ノードのネットワーク設定

各ノードのネットワーク設定は以下の通りです。

- コントローラーノード

インターフェース	eth0
IP アドレス	10.0.0.101
ネットマスク	255.255.255.0
ゲートウェイ	10.0.0.1
ネームサーバー	10.0.0.1

- コンピュートノード

インターフェース	eth0
IP アドレス	10.0.0.102
ネットマスク	255.255.255.0
ゲートウェイ	10.0.0.1
ネームサーバー	10.0.0.1

1.5 Ubuntu Server のインストール

インストール

2 台のサーバーに対し、Ubuntu Server をインストールします。要点は以下の通りです。

- 優先ネットワークインターフェースを eth0 に指定
    - インターネットへ接続するインターフェースは eth0 を使用するため、インストール中は eth0 を優先ネットワークとして指定します。
- OS は最小インストール
    - パッケージ選択では OpenSSH server のみ選択

第 1 章　構築する環境について

【インストール時の設定パラメータ例】

設定項目	設定例
初期起動時の Language	English
起動	Install Ubuntu Server
言語	English - English
地域の設定	other → Asia → Japan
地域の言語	United States - en_US.UTF-8
キーボードレイアウトの認識	No
キーボードの言語	Japanese → Japanese
優先する NIC	eth0: Ethernet
ホスト名	それぞれのノード名 (controller, compute)
ユーザ名とパスワード	フルネームで入力
アカウント名	ユーザ名のファーストネームで設定される
パスワード	任意のパスワード

パスワードを入力後、Weak password という警告が出た場合は Yes を選択するか、警告が出ないようなより複雑なパスワードを設定してください。

設定項目	設定例
ホームの暗号化	任意
タイムゾーン	Asia/Tokyo であることを確認
パーティション設定	Guided - use entire disk and set up LVM
パーティション選択	sda を選択
パーティション書き込み	Yes を選択
パーティションサイズ	デフォルトのまま
変更の書き込み	Yes を選択
HTTP proxy	環境に合わせて任意
アップグレード	No automatic updates を推奨
ソフトウェア	OpenSSH server のみ選択
GRUB	Yes を選択
インストール完了	Continue を選択

筆者注：
Ubuntu インストール時に選択した言語がインストール後も使われます。
Ubuntu Server で日本語の言語を設定した場合、標準出力や標準エラー出力が文字化けするなど様々な問題が起きますので、言語は英語を設定されることを推奨します。

プロキシーの設定

外部ネットワークとの接続にプロキシーの設定が必要な場合は、aptコマンドを使ってパッケージの照会やダウンロードを行うために次のような設定をする必要があります。

- システムのプロキシー設定

```
# vi /etc/environment
http_proxy="http://proxy.example.com:8080/"
https_proxy="https://proxy.example.com:8080/"
no_proxy=localhost,controller,compute
```

- APTのプロキシー設定

```
# vi /etc/apt/apt.conf
Acquire::http::proxy "http://proxy.example.com:8080/";
Acquire::https::proxy "https://proxy.example.com:8080/";
```

より詳細な情報は下記のサイトの情報を確認ください。

- https://help.ubuntu.com/community/AptGet/Howto
- http://gihyo.jp/admin/serial/01/ubuntu-recipe/0331

1.6 Ubuntu Serverへのログインとroot権限

Ubuntuはデフォルト設定でrootユーザーの利用を許可していないため、root権限が必要となる作業は以下のように行ってください。

- rootユーザーで直接ログインできないので、インストール時に作成したアカウントでログインする。
- root権限が必要な場合には、sudoコマンドを使用する。
- rootで連続して作業したい場合には、sudo -iコマンドでシェルを起動する。

1.7 設定ファイル等の記述について

- 設定ファイルは特別な記述が無い限り、必要な設定を抜粋したものです。
- 特に変更の必要がない設定項目は省略されています。

第 1 章　構築する環境について

- [見出し] が付いている場合、その見出しから次の見出しまでの間に設定を記述します。
- コメントアウトされていない設定項目が存在する場合には、値を変更してください。多くの設定項目は記述が存在しているため、エディタの検索機能で検索することをお勧めします。
- 特定のホストでコマンドを実行する場合はコマンドの冒頭にホスト名を記述しています。

【設定ファイルの記述例】

```
controller# vi /etc/glance/glance-api.conf ←コマンド冒頭にこのコマンドを実行するホストを記述

[DEFAULT]
debug=true                          ← コメントをはずす

[database]
#connection = sqlite:////var/lib/glance/glance.sqlite   ← 既存設定をコメントアウト
connection = mysql+pymysql://glance:password@controller/glance   ← 追記

[keystone_authtoken] ← 見出し
#auth_host = 127.0.0.1 ← 既存設定をコメントアウト
auth_host = controller ← 追記

auth_port = 35357
auth_protocol = http
auth_uri = http://controller:5000/v2.0 ← 追記
admin_tenant_name = service ← 変更
admin_user = glance ← 変更
admin_password = password ← 変更
```

第2章 OpenStack インストール前の設定

OpenStack パッケージのインストール前に各々のノードで以下の設定を行います。

- ネットワークデバイスの設定
- ホスト名と静的な名前解決の設定
- リポジトリーの設定とパッケージの更新
- Chrony サーバーのインストール（コントローラーノードのみ）
- Chrony クライアントのインストール
- Python 用 MySQL/MariaDB クライアントのインストール
- MariaDB のインストール（コントローラーノードのみ）
- RabbitMQ のインストール（コントローラーノードのみ）
- 環境変数設定ファイルの作成（コントローラーノードのみ）
- memcached のインストールと設定（コントローラーノードのみ）

2.1 ネットワークデバイスの設定

各ノードの/etc/network/interfaces を編集し、IP アドレスの設定を行います。

コントローラーノードの IP アドレスの設定

```
controller# vi /etc/network/interfaces
auto eth0
iface eth0 inet static
```

```
        address 10.0.0.101
        netmask 255.255.255.0
        gateway 10.0.0.1
        dns-nameservers 10.0.0.1
```

コンピュートノードの IP アドレスの設定

```
compute# vi /etc/network/interfaces
auto eth0
iface eth0 inet static
        address 10.0.0.102
        netmask 255.255.255.0
        gateway 10.0.0.1
        dns-nameservers 10.0.0.1
```

ネットワークの設定を反映

各ノードで変更した設定を反映させるため、ホストを再起動します。

```
# shutdown -r now
```

2.2　ホスト名と静的な名前解決の設定

　ホスト名でノードにアクセスするには DNS サーバーで名前解決する方法や hosts ファイルに書く方法が挙げられます。本書では各ノードの /etc/hosts に各ノードの IP アドレスとホスト名を記述して hosts ファイルを使って名前解決します。127.0.1.1 の行はコメントアウトします。

各ノードのホスト名の設定

　各ノードのホスト名を hostnamectl コマンドを使って設定します。反映させるためには一度ログインしなおす必要があります。

　（例）コントローラーノードの場合

```
controller# hostnamectl set-hostname controller
controller# cat /etc/hostname
controller
```

各ノードの/etc/hosts の設定

すべてのノードで 127.0.1.1 の行をコメントアウトします。またホスト名で名前引きできるように設定します。

（例）コントローラーノードの場合

```
controller# vi /etc/hosts
127.0.0.1 localhost
#127.0.1.1 controller  ← 既存設定をコメントアウト
#ext
10.0.0.101 controller
10.0.0.102 compute
```

2.3　リポジトリーの設定とパッケージの更新

コントローラーノードとコンピュートノードで以下のコマンドを実行し、Mitaka 向け Ubuntu Cloud Archive リポジトリーを登録します。

```
# apt-get install software-properties-common
# add-apt-repository cloud-archive:mitaka

 Ubuntu Cloud Archive for OpenStack Mitaka
 More info: https://wiki.ubuntu.com/ServerTeam/CloudArchive

Press [ENTER] to continue or ctrl-c to cancel adding it    ← Enter キーを押す
...
Importing ubuntu-cloud.archive.canonical.com keyring

OK

Processing ubuntu-cloud.archive.canonical.com removal keyring

gpg: /etc/apt/trustdb.gpg: trustdb created

OK
```

　各ノードのシステムをアップデートします。Ubuntu ではパッケージのインストールやアップデートの際にまず apt-get update を実行してリポジトリーの情報の更新が必要です。そのあと apt-get -y dist-upgrade でアップグレードを行います。カーネルの更新があった場合は再起動してください。

　なお、apt-get update は頻繁に実行する必要はありません。日をまたいで作業する際や、コマンドを実行しない場合にパッケージ更新やパッケージのインストールでエラーが出る場合は実行してください。以降の手順では apt-get update を省略します。

```
# apt-get update && apt-get dist-upgrade
```

2.4 OpenStackクライアントとMariaDBクライアントのインストール

コントローラーノードでOpenStackクライアントとPython用のMySQL/MariaDBクライアントをインストールします。依存するパッケージは全てインストールします。

```
controller# apt-get install python-openstackclient python-pymysql
```

2.5 時刻同期サーバーのインストールと設定

時刻同期サーバーChronyの設定

各ノードで時刻を正確にするために時刻同期サーバーのChronyをインストールします。

```
# apt-get install chrony
```

コントローラーノードの時刻同期サーバーの設定

コントローラーノードで公開NTPサーバーと同期するNTPサーバーを構築します。適切な公開NTPサーバー (ex.ntp.nict.jp etc..) を指定します。ネットワーク内にNTPサーバーがある場合はそのサーバーを指定します。設定を変更した場合はNTPサービスを再起動します。

```
controller# service chrony restart
```

その他ノードの時刻同期サーバーの設定

コンピュートノードでコントローラーノードと同期するNTPサーバーを構築します。

```
compute# vi /etc/chrony/chrony.conf

#server 0.debian.pool.ntp.org offline minpoll 8 #デフォルト設定はコメントアウト
#server 1.debian.pool.ntp.org offline minpoll 8
#server 2.debian.pool.ntp.org offline minpoll 8
#server 3.debian.pool.ntp.org offline minpoll 8

server controller iburst
```

設定を適用するため、NTP サービスを再起動します。

```
compute# service chrony restart
```

NTPサーバーの動作確認

構築した環境でコマンドを実行して、各 NTP サーバーが同期していることを確認します。

- 公開 NTP サーバーと同期しているコントローラーノード

```
controller# chronyc sources
chronyc sources
210 Number of sources = 4
MS Name/IP address          Stratum Poll Reach LastRx Last sample
===============================================================================
^* chobi.paina.jp                 2    8    17      1    +47us[ -312us] +/-   13ms
^- v157-7-235-92.z1d6.static      2    8    17      1  +1235us[+1235us] +/-   45ms
^- edm.butyshop.com               3    8    17      0  -2483us[-2483us] +/-   82ms
^- y.ns.gin.ntt.net               2    8    17      0  +1275us[+1275us] +/-   35ms
```

- コントローラーノードと同期しているその他ノード

```
compute# chronyc sources
210 Number of sources = 1
MS Name/IP address          Stratum Poll Reach LastRx Last sample
===============================================================================
^* controller                     3    6    77     25   -509us[-1484us] +/-   13ms
```

2.6 MariaDBのインストール

コントローラーノードにデータベースサーバーの MariaDB をインストールします。

パッケージのインストール

apt-get コマンドで mariadb-server パッケージをインストールします。

```
controller# apt-get install mariadb-server
```

インストール中にパスワードの入力を要求されますので、MariaDB の root ユーザーに対するパスワードを設定します。本書ではパスワードとして「password」を設定します。

MariaDBの設定を変更

OpenStack 用の MariaDB 設定ファイル /etc/mysql/conf.d/openstack.cnf を作成し、以下を設定します。

- バインドアドレス：eth0 へ割り当てた IP アドレス
- 文字コード：UTF-8

別のノードから MariaDB へアクセスできるようにするためバインドアドレスを変更します。加えて使用する文字コードを utf8 に変更します。

※文字コードを utf8 に変更しないと OpenStack モジュールとデータベース間の通信でエラーが発生します。

```
controller# vi /etc/mysql/conf.d/openstack.cnf

[mysqld]
bind-address = 10.0.0.101                    ← controller の IP アドレス
default-storage-engine = innodb
innodb_file_per_table
collation-server = utf8_general_ci
character-set-server = utf8
```

MariaDB サービスの再起動

変更した設定を反映させるため MariaDB のサービスを再起動します。

```
controller# service mysql restart
```

MariaDB クライアントのインストール

コンピュートノードに MariaDB クライアントをインストールします。コントローラーノードでインストール済みの MariaDB と同様のバージョンを指定します。

```
compute# apt-get install mariadb-client-5.5 mariadb-client-core-5.5
```

2.7　RabbitMQのインストール

OpenStack は、オペレーションやステータス情報を各サービス間で連携するためにメッセージブローカーを使用しています。OpenStack では RabbitMQ、ZeroMQ など複数のメッセージブローカーサービスに対応しています。本書では RabbitMQ をインストールする例を説明し

ます。

パッケージのインストール

コントローラーノードに rabbitmq-server パッケージをインストールします。

```
controller# apt-get install rabbitmq-server
```

openstack ユーザーの作成とパーミッションの設定

RabbitMQ にアクセスするためのユーザーとして openstack ユーザーを作成し、必要なパーミッションを設定します。以下コマンドは RabbitMQ のパスワードを password に設定する例です。

```
controller# rabbitmqctl add_user openstack password
controller# rabbitmqctl set_permissions openstack ".*" ".*" ".*"
```

待ち受け IP アドレス・ポートとセキュリティ設定の変更

以下の設定ファイルを作成し、RabbitMQ の待ち受けポートと IP アドレスを設定します。

- 待ち受け設定の追加

```
controller# vi /etc/rabbitmq/rabbitmq-env.conf
RABBITMQ_NODE_IP_ADDRESS=10.0.0.101     ← controller の IP アドレス
RABBITMQ_NODE_PORT=5672
HOSTNAME=controller
```

RabbitMQ サービス再起動と確認

- ログの確認

メッセージブローカーサービスが正常に動いていないと、OpenStack の各コンポーネントは正常に動きません。RabbitMQ サービスの再起動と動作確認を行い、確実に動作していることを確認します。

```
controller# service rabbitmq-server restart
controller# tailf /var/log/rabbitmq/rabbit@controller.log
...
 =INFO REPORT==== 16-May-2016::13:41:30 ===

started TCP Listener on 10.0.0.101:5672   ←待受 IP とポートを確認
```

```
=INFO REPORT==== 16-May-2016::13:41:30 ===
Server startup complete; 0 plugins started.
```

※新たなエラーが表示されなければ問題ありません。

2.8　環境変数設定ファイルの作成

admin 環境変数設定ファイルの作成

admin ユーザー用の環境変数設定ファイルを作成します。

```
controller# vi ~/admin-openrc.sh
export OS_PROJECT_DOMAIN_NAME=default
export OS_USER_DOMAIN_NAME=default
export OS_PROJECT_NAME=admin
export OS_USERNAME=admin
export OS_PASSWORD=password
export OS_AUTH_URL=http://controller:35357/v3
export OS_IDENTITY_API_VERSION=3
export OS_IMAGE_API_VERSION=2
export PS1='\u@\h \W(admin)\$ '
```

demo 環境変数設定ファイルの作成

demo ユーザー用の環境変数設定ファイルを作成します。

```
controller# vi ~/demo-openrc.sh
export OS_PROJECT_DOMAIN_NAME=default
export OS_USER_DOMAIN_NAME=default
export OS_PROJECT_NAME=demo
export OS_USERNAME=demo
export OS_PASSWORD=password
export OS_AUTH_URL=http://controller:5000/v3
export OS_IDENTITY_API_VERSION=3
export OS_IMAGE_API_VERSION=2
export PS1='\u@\h \W(demo)\$ '
```

2.9　memcached のインストールと設定

コントローラーノードで認証機構のトークンをキャッシュするための用途で memcached をインストールします。

2.9 memcached のインストールと設定

```
controller# apt-get install memcached python-memcache
```

設定ファイル /etc/memcached.conf を変更します。既存の行 -l 127.0.0.1 を controller の IP アドレスに変更します。

```
controller# vi /etc/memcached.conf
-l 10.0.0.101     ← controller の IP アドレス
```

memcached サービスを再起動します。

```
controller# service memcached restart
```

第3章 Keystoneのインストールと設定（コントローラーノード）

各サービス間の連携時に使用する認証サービスKeystoneのインストールと設定を行います。

3.1 データベースを作成

MariaDBにKeystoneで使用するデータベースを作成しアクセス権を付与します。

```
controller# mysql -u root -p << EOF
CREATE DATABASE keystone;
GRANT ALL PRIVILEGES ON keystone.* TO 'keystone'@'localhost' \
IDENTIFIED BY 'password';
GRANT ALL PRIVILEGES ON keystone.* TO 'keystone'@'%' \
IDENTIFIED BY 'password';
EOF
Enter password: ← MariaDBのrootパスワードpasswordを入力
```

3.2 データベースの確認

MariaDBにkeystoneユーザーでログインしデータベースの閲覧が可能であることを確認します。

```
controller# mysql -u keystone -p
Enter password:  ← MariaDBのkeystoneパスワードpasswordを入力
...
Type 'help;' or '\h' for help. Type '\c' to clear the current input statement.

MariaDB [(none)]> show databases;
+--------------------+
```

```
| Database           |
+--------------------+
| information_schema |
| keystone           |
+--------------------+
2 rows in set (0.00 sec)
```

3.3　admin_tokenの決定

Keystone の admin_token に設定する 10 桁の文字列を次のコマンドで決定します。出力結果はランダム英数字になります。

```
controller# openssl rand -hex 10
45742a05a541f26ddee8
```

3.4　パッケージのインストール

Keystone のインストール後、サービスが自動的に起動しないように以下のコマンドを実行し無効化します。

```
controller# echo "manual" > /etc/init/keystone.override
```

apt-get コマンドで keystone および必要なパッケージをインストールします。

```
controller# apt-get install keystone apache2 libapache2-mod-wsgi
```

3.5　Keystoneの設定変更

keystone の設定ファイルを変更します。

```
controller# vi /etc/keystone/keystone.conf

[DEFAULT]
admin_token = 45742a05a541f26ddee8     ← 追記 (3-3 で出力されたキーを入力)
log_dir = /var/log/keystone            ← 設定されていることを確認
...
[database]
#connection = sqlite:////var/lib/keystone/keystone.db     ← 既存設定をコメントアウト
connection = mysql+pymysql://keystone:password@controller/keystone  ← 追記
...
```

```
[token]
...
provider = fernet          ← 追記
```

次のコマンドで正しく設定を行ったか確認します。

```
controller# less /etc/keystone/keystone.conf | grep -v "^\s*$" | grep -v "^\s*#"
```

3.6 データベースに展開

```
controller# su -s /bin/sh -c "keystone-manage db_sync" keystone
```

3.7 Fernet キーの初期化

keystone-manage コマンドで Fernet キーを初期化します。

```
controller# keystone-manage fernet_setup --keystone-user keystone --keystone-group keystone
```

3.8 Apache Web サーバーの設定

- コントローラーノードの /etc/apache2/apache2.conf の ServerName にコントローラーノードのホスト名を設定します。

```
# Global configuration
#
ServerName controller
...
```

- コントローラーノードで/etc/apache2/sites-available/wsgi-keystone.conf を作成し、次の内容を記述します。

```
Listen 5000
Listen 35357

<VirtualHost *:5000>
    WSGIDaemonProcess keystone-public processes=5 threads=1 user=keystone group=keystone display-name=%{GROUP}
```

```
    WSGIProcessGroup keystone-public
    WSGIScriptAlias / /usr/bin/keystone-wsgi-public
    WSGIApplicationGroup %{GLOBAL}
    WSGIPassAuthorization On
    ErrorLogFormat "%{cu}t %M"
    ErrorLog /var/log/apache2/keystone.log
    CustomLog /var/log/apache2/keystone_access.log combined

    <Directory /usr/bin>
        Require all granted
    </Directory>
</VirtualHost>

<VirtualHost *:35357>
    WSGIDaemonProcess keystone-admin processes=5 threads=1 user=keystone
group=keystone display-name=%{GROUP}
    WSGIProcessGroup keystone-admin
    WSGIScriptAlias / /usr/bin/keystone-wsgi-admin
    WSGIApplicationGroup %{GLOBAL}
    WSGIPassAuthorization On
    ErrorLogFormat "%{cu}t %M"
    ErrorLog /var/log/apache2/keystone.log
    CustomLog /var/log/apache2/keystone_access.log combined

    <Directory /usr/bin>
        Require all granted
    </Directory>
</VirtualHost>
```

- keystone サービスの VirtualHost を有効化します。

```
controller# ln -s /etc/apache2/sites-available/wsgi-keystone.conf
/etc/apache2/sites-enabled
```

3.9　サービスの再起動と不要DBの削除

- Apache Web サーバーを再起動します。

```
controller# service apache2 restart
```

- パッケージのインストール時に作成される不要なSQLiteファイルを削除します。

```
controller# rm /var/lib/keystone/keystone.db
```

3.10　サービスとAPIエンドポイントの作成

以下コマンドでサービスとAPIエンドポイントを設定します。

- 環境変数の設定

```
controller# export OS_TOKEN=45742a05a541f26ddee8    ← 追記 (3-5 で設定したキーを入力)
controller# export OS_URL=http://controller:35357/v3
controller# export OS_IDENTITY_API_VERSION=3
```

- サービスの作成

```
controller# openstack service create \
 --name keystone --description "OpenStack Identity" identity

+-------------+----------------------------------+
| Field       | Value                            |
+-------------+----------------------------------+
| description | OpenStack Identity               |
| enabled     | True                             |
| id          | e56875264dad4553bcd5682f9ecc15b9 |
| name        | keystone                         |
| type        | identity                         |
+-------------+----------------------------------+
```

- APIエンドポイントの作成

```
controller# openstack endpoint create --region RegionOne \
  identity public http://controller:5000/v3

controller# openstack endpoint create --region RegionOne \
  identity internal http://controller:5000/v3

controller# openstack endpoint create --region RegionOne \
  identity admin http://controller:35357/v3
```

第 3 章　Keystone のインストールと設定（コントローラーノード）

3.11　プロジェクトとユーザー、ロールの作成

以下コマンドで認証情報（プロジェクト・ユーザー・ロール）を設定します。

- default ドメインの作成

```
controller# openstack domain create --description "Default Domain" default
```

- admin プロジェクトの作成

```
controller# openstack project create --domain default \
  --description "Admin Project" admin
```

- admin ユーザーの作成

```
controller# openstack user create --domain default --password-prompt admin
User Password: password    #admin ユーザーのパスワードを設定 (本書は password を設定)
Repeat User Password: password
+-----------+----------------------------------+
| Field     | Value                            |
+-----------+----------------------------------+
| domain_id | c12fc3119bb046a09ba53e0566004f76 |
| enabled   | True                             |
| id        | 59f4de8f7a7044d1b307dd3a89091bd6 |
| name      | admin                            |
+-----------+----------------------------------+
```

- admin ロールの作成

```
controller# openstack role create admin
+-------+----------------------------------+
| Field | Value                            |
+-------+----------------------------------+
| id    | e63b506dc78e45a6ab282dd68c7fc53f |
| name  | admin                            |
+-------+----------------------------------+
```

- admin ロールを admin プロジェクトと admin ユーザーに追加します。

```
controller# openstack role add --project admin --user admin admin
```

- サービスプロジェクトの作成

3.11 プロジェクトとユーザー、ロールの作成

```
controller# openstack project create --domain default \
  --description "Service Project" service
+-------------+----------------------------------+
| Field       | Value                            |
+-------------+----------------------------------+
| description | Service Project                  |
| domain_id   | c12fc3119bb046a09ba53e0566004f76 |
| enabled     | True                             |
| id          | a85f5704eb5e485d8de664132a3954bd |
| is_domain   | False                            |
| name        | service                          |
| parent_id   | c12fc3119bb046a09ba53e0566004f76 |
+-------------+----------------------------------+
```

- demo プロジェクトの作成

```
controller# openstack project create --domain default \
  --description "Demo Project" demo
+-------------+----------------------------------+
| Field       | Value                            |
+-------------+----------------------------------+
| description | Demo Project                     |
| domain_id   | c12fc3119bb046a09ba53e0566004f76 |
| enabled     | True                             |
| id          | 4e00692779864ed185356843ebeb0e2f |
| is_domain   | False                            |
| name        | demo                             |
| parent_id   | c12fc3119bb046a09ba53e0566004f76 |
+-------------+----------------------------------+
```

- demo ユーザーの作成

```
controller# openstack user create --domain default \
  --password-prompt demo
User Password: password    #demo ユーザーのパスワードを設定 (本書は password を設定)
Repeat User Password: password
+-----------+----------------------------------+
| Field     | Value                            |
+-----------+----------------------------------+
| domain_id | c12fc3119bb046a09ba53e0566004f76 |
| enabled   | True                             |
| id        | a5fab4bc7a0645cd85e68af50c02d2fe |
| name      | demo                             |
+-----------+----------------------------------+
```

- user ロールの作成

第 3 章　Keystone のインストールと設定（コントローラーノード）

```
controller# openstack role create user
+-------+----------------------------------+
| Field | Value                            |
+-------+----------------------------------+
| id    | 57c726c5060648898c5dd693a9661aa2 |
| name  | user                             |
+-------+----------------------------------+
```

- user ロールを demo プロジェクトと demo ユーザーに追加します。

```
controller# openstack role add --project demo --user demo user
```

3.12　Keystone の動作確認

他のサービスをインストールする前に Keystone が正しく構築、設定されたか動作を検証します。

- セキュリティを確保するため、一時認証トークンメカニズムを無効化します。/etc/keystone/keystone-paste.ini を開き、[pipeline:public_api] と [pipeline:admin_api] と [pipeline:api_v3] セクション（訳者注:... の pipeline 行）から、admin_token_auth を削除します。

```
[pipeline:public_api]
pipeline = sizelimit url_normalize request_id build_auth_context token_auth
json_body ec2_extension user_crud_extension public_service
...
[pipeline:admin_api]
pipeline = sizelimit url_normalize request_id build_auth_context token_auth
json_body ec2_extension s3_extension crud_extension admin_service
...
[pipeline:api_v3]
pipeline = sizelimit url_normalize request_id build_auth_context token_auth
json_body ec2_extension_v3 s3_extension simple_cert_extension revoke_extension
federation_extension oauth1_extension endpoint_filter_extension
endpoint_policy_extension service_v3
```

- 一時的に設定した環境変数 OS_TOKEN と OS_URL を解除します。

```
controller# unset OS_TOKEN OS_URL
```

動作確認のため admin および demo テナントに対し認証トークンを要求します。admin、

demo ユーザーのパスワードを入力します。

- admin ユーザーとして認証トークンを要求します。

```
controller# openstack --os-auth-url http://controller:35357/v3 \
  --os-project-domain-name default --os-user-domain-name default \
  --os-project-name admin --os-username admin token issue
Password:
+------------+----------------------------------+
| Field      | Value                            |
+------------+----------------------------------+
| expires    | 2016-05-17T02:08:06.019234Z      |
| id         | gAAAAABXOm72ayy8PZc2gSZi5CdO...  |
| project_id | 360855029dbe44649a5dc862c5a3caf1 |
| user_id    | 59f4de8f7a7044d1b307dd3a89091bd6 |
+------------+----------------------------------+
```

正常な応答だと、/var/log/apache2/keystone_access.log に「"POST /v3/auth/tokens HTTP/1.1" 201」が記録されます。例えば故意に認証用パスワードを間違えた場合は「"POST /v3/auth/tokens HTTP/1.1" 401」が記録されます。正常な応答がない場合は /var/log/apache2/keystone_error.log を確認しましょう。

```
...
10.0.0.101 - - [16/May/2016:15:54:49 +0900] "GET /v3 HTTP/1.1" 200 556 "-"
"python-openstackclient keystoneauth1/2.4.0 python-requests/2.9.1 CPython/2.7.6"
10.0.0.101 - - [16/May/2016:15:54:49 +0900] "POST /v3/auth/tokens HTTP/1.1" 201 1608
"-" "python-openstackclient keystoneauth1/2.4.0 python-requests/2.9.1 CPython/2.7.6"
```

- demo ユーザーとして管理トークンを要求します。

```
controller# openstack --os-auth-url http://controller:5000/v3 \
  --os-project-domain-name default --os-user-domain-name default \
  --os-project-name demo --os-username demo token issue
Password:
+------------+----------------------------------+
| Field      | Value                            |
+------------+----------------------------------+
| expires    | 2016-05-17T02:35:04.100816Z      |
| id         | gAAAAABXOnVIT0AYCEj0J_uWhI86I... |
| project_id | 4e00692779864ed185356843ebeb0e2f |
| user_id    | a5fab4bc7a0645cd85e68af50c02d2fe |
+------------+----------------------------------+
```

第4章 Glance のインストールと設定

4.1 データベースを作成

MariaDB に glance で使用するデータベースを作成し、アクセス権を付与します。

```
controller# mysql -u root -p << EOF
CREATE DATABASE glance;
GRANT ALL PRIVILEGES ON glance.* TO 'glance'@'localhost' \
 IDENTIFIED BY 'password';
GRANT ALL PRIVILEGES ON glance.* TO 'glance'@'%' \
IDENTIFIED BY 'password';
EOF
Enter password: ← MariaDB の root パスワード password を入力
```

4.2 データベースの確認

MariaDB に glance ユーザーでログインし、データベースの閲覧が可能であることを確認します。

```
controller# mysql -u glance -p
Enter password: ← MariaDB の glance パスワード password を入力
...
Type 'help;' or '\h' for help. Type '\c' to clear the current input statement.

MariaDB [(none)]> show databases;
+--------------------+
| Database           |
+--------------------+
| information_schema |
```

```
| glance            |
+-------------------+
2 rows in set (0.00 sec)
```

4.3 ユーザー、サービス、APIエンドポイントの作成

以下のコマンドで認証情報を読み込み、ImageサービスとAPIエンドポイントを設定します。

- 環境変数ファイルの読み込み

admin-openrc.sh を読み込むと次のようにプロンプトが変化します。

```
controller# source admin-openrc.sh
controller ~(admin)#
```

- glance ユーザーの作成

```
controller# openstack user create --domain default --password-prompt glance
User Password: password    #glance ユーザーのパスワードを設定 (本書は password を設定)
Repeat User Password: password
+-----------+----------------------------------+
| Field     | Value                            |
+-----------+----------------------------------+
| domain_id | c12fc3119bb046a09ba53e0566004f76 |
| enabled   | True                             |
| id        | 76ef779bd772488ab6e8d2984d184f69 |
| name      | glance                           |
+-----------+----------------------------------+
```

- admin ロールを glance ユーザーと service プロジェクトに追加

```
controller# openstack role add --project service --user glance admin
```

- glance サービスの作成

```
controller# openstack service create --name glance \
--description "OpenStack Image service" image
+-------------+----------------------------------+
| Field       | Value                            |
+-------------+----------------------------------+
```

```
| description | OpenStack Image service         |
| enabled     | True                             |
| id          | 746ddec3462a4075845dfc76fb05f09f |
| name        | glance                           |
| type        | image                            |
+-------------+----------------------------------+
```

- API エンドポイントの作成

```
controller# openstack endpoint create --region RegionOne \
  image public http://controller:9292
controller# openstack endpoint create --region RegionOne \
  image internal http://controller:9292
controller# openstack endpoint create --region RegionOne \
  image admin http://controller:9292
```

4.4　Glanceのインストール

apt-get コマンドで glance パッケージをインストールします。

```
controller# apt-get install glance
```

4.5　Glanceの設定変更

Glance の設定を行います。glance-api.conf、glance-registry.conf ともに、[keystone_authtoken] に追記した設定以外のパラメーターはコメントアウトします。

```
controller# vi /etc/glance/glance-api.conf
...
[database]
#sqlite_db = /var/lib/glance/glance.sqlite               ← 既存設定をコメントアウト
connection = mysql+pymysql://glance:password@controller/glance   ← 追記
...
[glance_store]
...
stores = file,http
default_store = file
filesystem_store_datadir = /var/lib/glance/images/      ← 追記
...
[keystone_authtoken] （既存の設定はコメントアウトし、以下を追記）
...
auth_uri = http://controller:5000
```

```
auth_url = http://controller:35357
memcached_servers = controller:11211
auth_type = password
project_domain_name = default
user_domain_name = default
project_name = service
username = glance
password = password    ← glance ユーザーのパスワード
...
[paste_deploy]
...
flavor = keystone              ← 追記
```

次のコマンドで正しく設定を行ったか確認します。

```
controller# less /etc/glance/glance-api.conf | grep -v "^\s*$" | grep -v "^\s*#"
```

```
controller# vi /etc/glance/glance-registry.conf

[DEFAULT]
...
[database]
#sqlite_db = /var/lib/glance/glance.sqlite                    ← 既存設定をコメントアウト
connection = mysql+pymysql://glance:password@controller/glance   ← 追記

[keystone_authtoken]（既存の設定はコメントアウトし、以下を追記）
...
auth_uri = http://controller:5000
auth_url = http://controller:35357
memcached_servers = controller:11211
auth_type = password
project_domain_name = default
user_domain_name = default
project_name = service
username = glance
password = password        ← glance ユーザーのパスワード

...
[paste_deploy]
flavor = keystone              ← 追記
```

次のコマンドで正しく設定を行ったか確認します。

```
controller# less /etc/glance/glance-registry.conf | grep -v "^\s*$" | grep -v "^\s*#"
```

4.6　データベースに展開

次のコマンドで glance データベースのセットアップを行います。

```
controller# su -s /bin/sh -c "glance-manage db_sync" glance
```

※ 廃止予定に関するメッセージが出力されますが全て無視してください

4.7　Glance サービスの再起動

設定を反映させるため Glance サービスを再起動します。

```
controller# service glance-registry restart && service glance-api restart
```

4.8　ログの確認と使用しないデータベースファイルの削除

サービスの再起動後、ログを参照し Glance Registry と Glance API サービスでエラーが起きていないことを確認します。

```
controller# tailf /var/log/glance/glance-api.log
controller# tailf /var/log/glance/glance-registry.log
```

インストール直後は作られていない場合が多いですが、コマンドを実行して glance.sqlite を削除します。

```
controller# rm /var/lib/glance/glance.sqlite
```

4.9　イメージの取得と登録

Glance へインスタンス用の仮想マシンイメージを登録します。ここでは、OpenStack のテスト環境に役立つ軽量な Linux イメージ CirrOS を登録します。

イメージの取得

CirrOS の Web サイトより仮想マシンイメージをダウンロードします。

```
controller# wget http://download.cirros-cloud.net/0.3.4/cirros-0.3.4-x86_64-disk.img
```

イメージを登録

ダウンロードした仮想マシンイメージを Glance に登録します。

```
controller# glance image-create --name "cirros-0.3.4-x86_64" \
 --file cirros-0.3.4-x86_64-disk.img --disk-format qcow2 \
 --container-format bare \
 --visibility public --progress
[=============================>] 100%
+------------------+--------------------------------------+
| Property         | Value                                |
+------------------+--------------------------------------+
| checksum         | ee1eca47dc88f4879d8a229cc70a07c6     |
| container_format | bare                                 |
| created_at       | 2016-05-17T04:35:24Z                 |
| disk_format      | qcow2                                |
| id               | 12d67808-c7d2-44b4-8bc6-b0ced1878f06 |
| min_disk         | 0                                    |
| min_ram          | 0                                    |
| name             | cirros-0.3.4-x86_64                  |
| owner            | 360855029dbe44649a5dc862c5a3caf1     |
| protected        | False                                |
| size             | 13287936                             |
| status           | active                               |
| tags             | []                                   |
| updated_at       | 2016-05-17T04:35:25Z                 |
| virtual_size     | None                                 |
| visibility       | public                               |
+------------------+--------------------------------------+
```

イメージの登録を確認

仮想マシンイメージが正しく登録されたか確認します。

```
controller# openstack image list
+--------------------------------------+---------------------+--------+
| ID                                   | Name                | Status |
+--------------------------------------+---------------------+--------+
| 12d67808-c7d2-44b4-8bc6-b0ced1878f06 | cirros-0.3.4-x86_64 | active |
+--------------------------------------+---------------------+--------+
```

第5章 Novaのインストールと設定（コントローラーノード）

5.1 データベースを作成

MariaDBにデータベースnovaを作成します。

```
controller# mysql -u root -p << EOF
CREATE DATABASE nova_api;
CREATE DATABASE nova;
GRANT ALL PRIVILEGES ON nova_api.* TO 'nova'@'localhost' \
IDENTIFIED BY 'password';
GRANT ALL PRIVILEGES ON nova_api.* TO 'nova'@'%' \
IDENTIFIED BY 'password';
GRANT ALL PRIVILEGES ON nova.* TO 'nova'@'localhost' \
IDENTIFIED BY 'password';
GRANT ALL PRIVILEGES ON nova.* TO 'nova'@'%' \
IDENTIFIED BY 'password';
EOF
Enter password:            ← MariaDBのrootパスワードpasswordを入力
```

5.2 データベースの確認

MariaDBにnovaユーザーでログインし、データベースの閲覧が可能であることを確認します。

```
controller# mysql -u nova -p
Enter password: ← MariaDBのnovaパスワードpasswordを入力
...
Type 'help;' or '\h' for help. Type '\c' to clear the current input statement.
```

```
MariaDB [(none)]> show databases;
+--------------------+
| Database           |
+--------------------+
| information_schema |
| nova               |
| nova_api           |
+--------------------+
3 rows in set (0.00 sec)
```

5.3 ユーザーとサービス、APIエンドポイントの作成

以下コマンドで認証情報を読み込んだあと、サービスと API エンドポイントを設定します。

- 環境変数ファイルの読み込み

```
controller# source admin-openrc.sh
```

- nova ユーザーの作成

```
controller# openstack user create --domain default --password-prompt nova
User Password: password   #nova ユーザーのパスワードを設定 (本書は password を設定)
Repeat User Password: password
+-----------+----------------------------------+
| Field     | Value                            |
+-----------+----------------------------------+
| domain_id | c12fc3119bb046a09ba53e0566004f76 |
| enabled   | True                             |
| id        | e14bbac223e546559251f7e607693316 |
| name      | nova                             |
+-----------+----------------------------------+
```

- nova ユーザーを admin ロールに追加

```
controller# openstack role add --project service --user nova admin
```

- nova サービスの作成

```
controller# openstack service create --name nova \
  --description "OpenStack Compute" compute
+-------------+----------------------------------+
| Field       | Value                            |
+-------------+----------------------------------+
| description | OpenStack Compute                |
| enabled     | True                             |
| id          | 87d0a9fec6244c37902694b2ae3143a4 |
| name        | nova                             |
| type        | compute                          |
+-------------+----------------------------------+
```

- Compute サービスの API エンドポイントを作成

```
controller# openstack endpoint create --region RegionOne \
  compute public http://controller:8774/v2.1/%\(tenant_id\)s
controller# openstack endpoint create --region RegionOne \
  compute internal http://controller:8774/v2.1/%\(tenant_id\)s
controller# openstack endpoint create --region RegionOne \
  compute admin http://controller:8774/v2.1/%\(tenant_id\)s
```

5.4 パッケージのインストール

apt-get コマンドで Nova 関連のパッケージをインストールします。

```
controller# apt-get install nova-api nova-conductor nova-consoleauth \
 nova-novncproxy nova-scheduler
```

5.5 Novaの設定変更

nova.conf に下記の設定を追記します。

```
controller# vi /etc/nova/nova.conf

[DEFAULT]
dhcpbridge_flagfile=/etc/nova/nova.conf
dhcpbridge=/usr/bin/nova-dhcpbridge
logdir=/var/log/nova
state_path=/var/lib/nova
#lock_path=/var/lock/nova        ← コメントアウト
force_dhcp_release=True
```

```
libvirt_use_virtio_for_bridges=True
verbose=True
ec2_private_dns_show_ip=True
api_paste_config=/etc/nova/api-paste.ini
#enabled_apis=ec2,osapi_compute,metadata     ←コメントアウト
enabled_apis = osapi_compute,metadata        ←以下追記
rpc_backend = rabbit
auth_strategy = keystone

# コントローラーノードのIPアドレス:10.0.0.101
my_ip = 10.0.0.101                           ←追記

use_neutron = True

firewall_driver = nova.virt.firewall.NoopFirewallDriver

[vnc]
vncserver_listen = 10.0.0.101                        ←追記
vncserver_proxyclient_address = 10.0.0.101   ←自ホストを指定

(↓これ以下追記↓)

[api_database]
connection = mysql+pymysql://nova:password@controller/nova_api

[database]
connection = mysql+pymysql://nova:password@controller/nova

[oslo_messaging_rabbit]
rabbit_host = controller
rabbit_userid = openstack
rabbit_password = password

[keystone_authtoken]
auth_uri = http://controller:5000
auth_url = http://controller:35357
memcached_servers = controller:11211
auth_type = password
project_domain_name = default
user_domain_name = default
project_name = service
username = nova
password = password     ← novaユーザーのパスワード

[glance]
api_servers = http://controller:9292

[oslo_concurrency]
lock_path = /var/lib/nova/tmp
```

次のコマンドで正しく設定を行ったか確認します。

```
controller# less /etc/nova/nova.conf | grep -v "^\s*$" | grep -v "^\s*#"
```

5.6 データベースに展開

次のコマンドで nova データベースのセットアップを行います。

```
controller# su -s /bin/sh -c "nova-manage api_db sync" nova
controller# su -s /bin/sh -c "nova-manage db sync" nova
```

5.7 Nova サービスの再起動

設定を反映させるため、Nova のサービスを再起動します。

```
controller# service nova-api restart && service nova-consoleauth restart && service nova-scheduler restart && \
service nova-conductor restart && service nova-novncproxy restart
```

5.8 不要なデータベースファイルの削除

データベースは MariaDB を使用するため、不要な SQLite ファイルを削除します。

```
controller# rm /var/lib/nova/nova.sqlite
```

5.9 Glance との通信確認

Nova のコマンドラインインターフェースで Glance と通信して Glance と相互に通信できているかを確認します。

```
controller# nova image-list
+--------------------------------------+---------------------+--------+--------+
| ID                                   | Name                | Status | Server |
+--------------------------------------+---------------------+--------+--------+
| 12d67808-c7d2-44b4-8bc6-b0ced1878f06 | cirros-0.3.4-x86_64 | ACTIVE |        |
+--------------------------------------+---------------------+--------+--------+
```

※ Glance に登録した CirrOS イメージが表示できれば問題ありません。

第6章 Nova Computeのインストールと設定（コンピュートノード）

ここまでコントローラーノードの環境構築を行ってきましたが、ここでコンピュートノードに切り替えて設定を行います。

6.1 パッケージのインストール

```
compute# apt-get install nova-compute
```

6.2 Novaの設定を変更

novaの設定ファイルを変更します。

```
compute# vi /etc/nova/nova.conf

[DEFAULT]
dhcpbridge_flagfile=/etc/nova/nova.conf
dhcpbridge=/usr/bin/nova-dhcpbridge
logdir=/var/log/nova
state_path=/var/lib/nova
#lock_path=/var/lock/nova      ←コメントアウト
force_dhcp_release=True
libvirt_use_virtio_for_bridges=True
verbose=True
ec2_private_dns_show_ip=True
api_paste_config=/etc/nova/api-paste.ini
enabled_apis=ec2,osapi_compute,metadata
rpc_backend = rabbit
auth_strategy = keystone
```

第6章　Nova Compute のインストールと設定（コンピュートノード）

```
my_ip = 10.0.0.102          ← IP アドレスで指定

use_neutron = True
firewall_driver = nova.virt.firewall.NoopFirewallDriver

[vnc]
enabled = True
vncserver_listen = 0.0.0.0
vncserver_proxyclient_address = 10.0.0.102    ←自ホストを指定
novncproxy_base_url = http://10.0.0.101:6080/vnc_auto.html  ← novnc ホストを指定
vnc_keymap = ja                              ←日本語キーボードの設定

[oslo_messaging_rabbit]
rabbit_host = controller
rabbit_userid = openstack
rabbit_password = password

[keystone_authtoken]
auth_uri = http://controller:5000
auth_url = http://controller:35357
memcached_servers = controller:11211
auth_type = password
project_domain_name = default
user_domain_name = default
project_name = service
username = nova
password = password     ← nova ユーザーのパスワード

[glance]
api_servers = http://controller:9292

[oslo_concurrency]
lock_path = /var/lib/nova/tmp
```

次のコマンドで正しく設定を行ったか確認します。

```
compute# less /etc/nova/nova.conf | grep -v "^\s*$" | grep -v "^\s*#"
```

まず次のコマンドを実行し、コンピュートノードで Linux KVM が動作可能であることを確認します。コマンド結果が 1 以上の場合は、CPU 仮想化支援機能がサポートされています。もしこのコマンド結果が 0 の場合は仮想化支援機能がサポートされていないか、設定が有効化されていないので、libvirt で KVM の代わりに QEMU を使用します。後述の /etc/nova/nova-compute.conf の設定で virt_type = qemu を設定します。

```
compute# cat /proc/cpuinfo |egrep 'vmx|svm'|wc -l
4
```

VMX もしくは SVM 対応 CPU の場合は virt_type = kvm と設定することにより、仮想化

部分のパフォーマンスが向上します。

```
compute# vi /etc/nova/nova-compute.conf

[libvirt]
...
virt_type = kvm
```

6.3 Nova コンピュートサービスの再起動

設定を反映させるため、Nova-Compute サービスを再起動します。

```
compute# service nova-compute restart
```

6.4 コントローラーノードとの疎通確認

疎通確認はコントローラーノード上にて、admin 環境変数設定ファイルを読み込んで行います。

```
controller# source admin-openrc.sh
```

ホストリストの確認

コントローラーノードとコンピュートノードが相互に接続できているか確認します。もし、State が XXX になっているサービスがある場合は、該当のサービスのログを確認して対処してください。

```
controller# openstack compute service list -c Binary -c Host -c State
+------------------+------------+-------+
| Binary           | Host       | State |
+------------------+------------+-------+
| nova-consoleauth | controller | up    |
| nova-scheduler   | controller | up    |   ← Nova のステータスを確認
| nova-conductor   | controller | up    |
| nova-compute     | compute    | up    |
+------------------+------------+-------+
```

※一覧に compute が表示されていれば問題ありません。State が up でないサービスがある場合は -c オプションを外して確認します。

第 6 章　Nova Compute のインストールと設定（コンピュートノード）

ハイパーバイザの確認

コントローラーノードよりコンピュートノードのハイパーバイザが取得可能か確認します。

```
controller# openstack hypervisor list
+----+---------------------+
| ID | Hypervisor Hostname |
+----+---------------------+
|  1 | compute             |
+----+---------------------+
```

※ Hypervisor 一覧に compute が表示されていれば問題ありません。

第7章 Neutron のインストール・設定（コントローラーノード）

7.1 データベースを作成

MariaDB にデータベース neutron を作成し、アクセス権を付与します。

```
controller# mysql -u root -p << EOF
CREATE DATABASE neutron;
GRANT ALL PRIVILEGES ON neutron.* TO 'neutron'@'localhost' \
  IDENTIFIED BY 'password';
GRANT ALL PRIVILEGES ON neutron.* TO 'neutron'@'%' \
  IDENTIFIED BY 'password';
EOF
Enter password: ← MariaDB の root パスワード password を入力
```

7.2 データベースの確認

MariaDB に neutron ユーザーでログインし、データベースの閲覧が可能か確認します。

```
controller# mysql -u neutron -p
Enter password: ← MariaDB の neutron パスワード password を入力
...
Type 'help;' or '\h' for help. Type '\c' to clear the current input statement.

MariaDB [(none)]> show databases;
+--------------------+
| Database           |
+--------------------+
| information_schema |
| neutron            |
+--------------------+
```

```
2 rows in set (0.00 sec)
```

※ユーザー neutron でログイン可能でデータベースが閲覧可能なら問題ありません。

7.3 neutron ユーザーとサービス、API エンドポイントの作成

以下コマンドで認証情報を読み込んだあと、neutron サービスの作成と API エンドポイントを設定します。

- 環境変数ファイルの読み込み

```
controller# source admin-openrc.sh
```

- neutron ユーザーの作成

```
controller# openstack user create --domain default --password-prompt neutron
User Password: password   #neutron ユーザーのパスワードを設定 (本書は password を設定)
Repeat User Password: password
+-----------+----------------------------------+
| Field     | Value                            |
+-----------+----------------------------------+
| domain_id | c12fc3119bb046a09ba53e0566004f76 |
| enabled   | True                             |
| id        | 3ff553c90945412fbbdc5cc43110b308 |
| name      | neutron                          |
+-----------+----------------------------------+
```

- neutron ユーザーを admin ロールに追加

```
controller# openstack role add --project service --user neutron admin
```

- neutron サービスの作成

```
controller# openstack service create --name neutron --description "OpenStack Networking" network
+-------------+----------------------------------+
| Field       | Value                            |
+-------------+----------------------------------+
| description | OpenStack Networking             |
| enabled     | True                             |
```

```
| id           | 45e2fd18f80f465fabaf5dfa52742253 |
| name         | neutron                          |
| type         | network                          |
+--------------+----------------------------------+
```

- neutron サービスの API エンドポイントを作成

```
controller# openstack endpoint create --region RegionOne \
  network public http://controller:9696
controller# openstack endpoint create --region RegionOne \
  network internal http://controller:9696
controller# openstack endpoint create --region RegionOne \
  network admin http://controller:9696
```

7.4　パッケージのインストール

　本書ではネットワークの構成は公式マニュアルの「Networking Option 2: Self-service networks」の方法で構築する例を示します。

```
controller# apt-get install neutron-server neutron-plugin-ml2 \
  neutron-linuxbridge-agent neutron-l3-agent neutron-dhcp-agent \
  neutron-metadata-agent
```

7.5　Neutron コンポーネントの設定

- Neutron サーバーの設定

```
controller# vi /etc/neutron/neutron.conf

[DEFAULT]
...
core_plugin = ml2                    ←確認
service_plugins = router             ←追記
allow_overlapping_ips = True         ←追記

rpc_backend = rabbit                 ←コメントをはずす
auth_strategy = keystone             ←コメントをはずす
notify_nova_on_port_status_changes = True    ←コメントをはずす
notify_nova_on_port_data_changes = True      ←コメントをはずす

[database]
#connection = sqlite:////var/lib/neutron/neutron.sqlite   ←既存設定をコメントアウト
```

第 7 章 Neutron のインストール・設定（コントローラーノード）

```
connection = mysql+pymysql://neutron:password@controller/neutron   ←追記

[keystone_authtoken]（既存の設定はコメントアウトし、以下を追記）
...
auth_uri = http://controller:5000
auth_url = http://controller:35357
memcached_servers = controller:11211
auth_type = password
project_domain_name = default
user_domain_name = default
project_name = service
username = neutron
password = password   ← neutron ユーザーのパスワード

[nova]（以下末尾に追記）
...
auth_url = http://controller:35357
auth_type = password
project_domain_name = default
user_domain_name = default
region_name = RegionOne
project_name = service
username = nova
password = password   ← nova ユーザーのパスワード

[oslo_messaging_rabbit]（以下追記）
...
rabbit_host = controller
rabbit_userid = openstack
rabbit_password = password
```

[keystone_authtoken] セクションは追記した設定以外は取り除くかコメントアウトしてください。

次のコマンドで正しく設定を行ったか確認します。

```
controller# less /etc/neutron/neutron.conf | grep -v "^\s*$" | grep -v "^\s*#"
```

- ML2 プラグインの設定

```
controller# vi /etc/neutron/plugins/ml2/ml2_conf.ini

[ml2]
...
type_drivers = flat,vlan,vxlan              ←追記
tenant_network_types = vxlan                ←追記
mechanism_drivers = linuxbridge,l2population   ←追記 ※
extension_drivers = port_security           ←追記

[ml2_type_flat]
```

7.5 Neutron コンポーネントの設定

```
...
flat_networks = provider              ←追記

[ml2_type_vxlan]
...
vni_ranges = 1:1000                   ←追記

[securitygroup]
...
enable_ipset = True                   ←コメントをはずす
```

次のコマンドで正しく設定を行ったか確認します。

```
controller# less /etc/neutron/plugins/ml2/ml2_conf.ini | grep -v "^\s*$" | grep -v "^\s*#"
```

- Linux ブリッジエージェントの設定

パブリックネットワークに接続している側の NIC を指定します。本書では eth0 を指定します。

```
controller# vi /etc/neutron/plugins/ml2/linuxbridge_agent.ini

[linux_bridge]
physical_interface_mappings = provider:eth0 ←追記
```

local_ip は、先に physical_interface_mapping に設定した NIC 側の IP アドレスを設定します。

```
[vxlan]
enable_vxlan = True                   ←コメントをはずす
local_ip = 10.0.0.101                 ←追記
l2_population = True                  ←追記 ※
```

エージェントとセキュリティグループの設定を行います。

```
[securitygroup]
...
enable_security_group = True          ←コメントをはずす
firewall_driver = neutron.agent.linux.iptables_firewall.IptablesFirewallDriver
↑ 追記
```

次のコマンドで正しく設定を行ったか確認します。

```
controller# less /etc/neutron/plugins/ml2/linuxbridge_agent.ini | grep -v "^\s*$" | grep -v "^\s*#"
```

第 7 章　Neutron のインストール・設定（コントローラーノード）

※ ML2 プラグインの l2population について

　OpenStack Mitaka の公式手順書では、ML2 プラグインの mechanism_drivers として linuxbridge と l2population が指定されています。l2population が有効だとこれまでの動きと異なり、インスタンスが起動してもネットワークが有効化されるまで通信ができません。つまり Neutron ネットワークを構築してルーターのパブリック側の IP アドレス宛に Ping コマンドを実行して確認できても疎通ができません。このネットワーク有効化の有無についてメッセージキューサービスが監視して制御しています。

　従って OpenStack Mitaka では、これまでのバージョン以上にメッセージキューサービス（本例や公式手順ではしばしば RabbitMQ が使用されます）が確実に動作している必要があります。このような仕組みが導入されたのは、不要なパケットがネットワーク内に流れ続けないようにするためです。

　ただし、弊社で ESXi 仮想マシン環境に構築した OpenStack 環境において l2population が有効化されていると想定通り動かないという事象が発生することを確認してます。その他のハイパーバイザーでは確認していませんが、ネットワーク通信に支障が起きる場合は l2population をオフに設定すると改善される場合があります。修正箇所は次の通りです。

- controller の /etc/neutron/plugins/ml2/ml2_conf.ini の設定変更

```
[ml2]
...
mechanism_drivers = linuxbridge,l2population
↓
mechanism_drivers = linuxbridge
```

- controller & compute の /etc/neutron/plugins/ml2/linuxbridge_agent.ini の設定変更

```
[vxlan]
...
l2_population = true
↓
l2_population = false
```

　設定変更後は l2population の設定変更を反映させるため、controller と compute ノードの Neutron 関連サービスを再起動するか、システムを再起動してください。

- Layer-3 エージェントの設定

　external_network_bridge は単一のエージェントで複数の外部ネットワークを有効にするに

は値を指定してはならないため、値を空白にします。

```
controller# vi /etc/neutron/l3_agent.ini

[DEFAULT]    (最終行に以下を追記)
...
interface_driver = neutron.agent.linux.interface.BridgeInterfaceDriver
external_network_bridge =
```

- DHCP エージェントの設定

```
controller# vi /etc/neutron/dhcp_agent.ini

[DEFAULT]    (最終行に以下を追記)
...
interface_driver = neutron.agent.linux.interface.BridgeInterfaceDriver
dhcp_driver = neutron.agent.linux.dhcp.Dnsmasq
enable_isolated_metadata = True
```

- dnsmasq の設定

　一般的にデフォルトのイーサネットの MTU は 1500 に設定されています。通常の Ethernet フレームに VXLAN ヘッダが加算されるため、VXLAN を使う場合は少なくとも 50 バイト多い、1550 バイト以上の MTU が設定されていないと通信が不安定になったり、通信が不可能になる場合があります。

　これらはジャンボフレームを設定することで約 9000 バイトまでの MTU をサポートできるようになり対応可能ですが、ジャンボフレーム非対応のネットワーク機器を使う場合や、ネットワーク機器の設定を変更できない場合は VXLAN の 50 バイトのオーバーヘッドを考慮して 1450 バイト以内の MTU に設定する必要があります。

　これらの制約事項は OpenStack 環境でも同様で、インスタンスを起動する際に MTU 1450 を設定することで、この問題を回避可能です。

　この設定はインスタンス起動毎に UserData を使って設定することも可能ですが、次のように設定しておくと仮想 DHCP サーバーで MTU の自動設定を行うことができるので便利です。

- DHCP エージェントに dnsmasq の設定を追記

```
controller# vi /etc/neutron/dhcp_agent.ini

[DEFAULT]
...
```

```
dnsmasq_config_file = /etc/neutron/dnsmasq-neutron.conf   ←追記
```

- DHCP オプションの 26 番 (MTU) を定義

```
controller# vi /etc/neutron/dnsmasq-neutron.conf
dhcp-option-force=26,1450
```

- Metadata エージェントの設定

インスタンスのメタデータサービスを提供する Metadata エージェントを設定します。

```
controller# vi /etc/neutron/metadata_agent.ini
[DEFAULT]
nova_metadata_ip = controller   ← Metadata ホストを指定
metadata_proxy_shared_secret = METADATA_SECRET
```

　Metadata エージェントの metadata_proxy_shared_secret に指定する値と、次の手順で Nova に設定する metadata_proxy_shared_secret が同じ値になるように設定します。任意の値を設定すれば良いですが、思いつかない場合は次のように実行して生成した乱数を使うことも可能です。

```
controller# openssl rand -hex 10
```

次のコマンドで正しく設定を行ったか確認します。

```
controller# less /etc/neutron/metadata_agent.ini | grep -v "^\s*$" | grep -v "^\s*#"
```

7.6　Nova の設定を変更

　Nova の設定ファイルに Neutron の設定を追記します。

```
controller# vi /etc/nova/nova.conf
...
[neutron]
url = http://controller:9696
auth_url = http://controller:35357
auth_type = password
project_domain_name = default
user_domain_name = default
region_name = RegionOne
project_name = service
username = neutron
password = password     ← neutron ユーザーのパスワード
```

```
service_metadata_proxy = True
metadata_proxy_shared_secret = METADATA_SECRET
```

METADATA_SECRET は Metadata エージェントで指定した値に置き換えます。

次のコマンドで正しく設定を行ったか確認します。

```
controller# less /etc/nova/nova.conf | grep -v "^\s*$" | grep -v "^\s*#"
```

7.7 データベースに展開

コマンドを実行して、エラーが発生せずに完了することを確認します。

```
controller# su -s /bin/sh -c "neutron-db-manage \
 --config-file /etc/neutron/neutron.conf \
 --config-file /etc/neutron/plugins/ml2/ml2_conf.ini upgrade head" neutron

No handlers could be found for logger "neutron.quota"
INFO  [alembic.runtime.migration] Context impl MySQLImpl.
INFO  [alembic.runtime.migration] Will assume non-transactional DDL.
...
INFO  [alembic.runtime.migration] Running upgrade kilo -> c40fbb377ad, Initial Liberty no-op script.
INFO  [alembic.runtime.migration] Running upgrade c40fbb377ad -> 4b47ea298795, add reject rule
  OK
```

7.8 コントローラーノードの Neutron と関連サービスの再起動

設定を反映させるため、コントローラーノードの関連サービスを再起動します。まず Nova API サービスを再起動します。

```
controller# service nova-api restart
```

次に Neutron 関連サービスを再起動します。

```
controller# service neutron-server restart && \
 service neutron-linuxbridge-agent restart && \
 service neutron-dhcp-agent restart && \
 service neutron-metadata-agent restart && \
 service neutron-l3-agent restart
```

7.9 ログの確認

ログを確認して、エラーが出力されていないことを確認します。

```
controller# tailf /var/log/nova/nova-api.log
controller# tailf /var/log/neutron/neutron-server.log
controller# tailf /var/log/neutron/neutron-metadata-agent.log
controller# tailf /var/log/neutron/neutron-linuxbridge-agent.log
```

7.10 使用しないデータベースファイルを削除

```
controller# rm /var/lib/neutron/neutron.sqlite
```

第8章 Neutronのインストール・設定（コンピュートノード）

次にコンピュートノードの設定を行います。

8.1 パッケージのインストール

```
compute# apt-get install neutron-linuxbridge-agent
```

8.2 設定の変更

- Neutronの設定

```
compute# vi /etc/neutron/neutron.conf

[DEFAULT]
...
rpc_backend = rabbit                    ←コメントをはずす
auth_strategy = keystone                ←コメントをはずす

[keystone_authtoken]（既存の設定はコメントアウトし、以下を追記）
...
auth_uri = http://controller:5000
auth_url = http://controller:35357
memcached_servers = controller:11211
auth_type = password
project_domain_name = default
user_domain_name = default
project_name = service
username = neutron
```

第 8 章　Neutron のインストール・設定（コンピュートノード）

```
password = password    ← neutron ユーザーのパスワード
[database]
...
# connection = sqlite:////var/lib/neutron/neutron.sqlite    ← コメントアウト
[oslo_messaging_rabbit]
...
rabbit_host = controller          ←追記
rabbit_userid = openstack         ←追記
rabbit_password = password        ←追記
```

　本書の構成では、コンピュートノードの neutron.conf にはデータベースの指定は不要です。次のコマンドで正しく設定を行ったか確認します。

```
compute# less /etc/neutron/neutron.conf | grep -v "^\s*$" | grep -v "^\s*#"
```

- Linux ブリッジエージェントの設定

　physical_interface_mappings にはパブリック側のネットワークに接続しているインターフェイスを指定します。本書では eth0 を指定します。local_ip にはパブリック側に接続している NIC に設定している IP アドレスを指定します。

　追記と書かれていない項目は設定があればコメントをはずして設定を変更、なければ追記してください。

```
compute# vi /etc/neutron/plugins/ml2/linuxbridge_agent.ini

[linux_bridge]
physical_interface_mappings = provider:eth0

[vxlan]
enable_vxlan = True
local_ip = 10.0.0.102
l2_population = True    (※ 1)

[securitygroup]
...
enable_security_group = True
firewall_driver = neutron.agent.linux.iptables_firewall.IptablesFirewallDriver
 ↑追記
```

　※ 1 ML2 プラグインの設定 /etc/neutron/plugins/ml2/ml2_conf.ini で l2population を使用しない場合は l2_population = False を設定し無効化します。

　次のコマンドで正しく設定を行ったか確認します。

```
compute# less /etc/neutron/plugins/ml2/linuxbridge_agent.ini | grep -v "^\s*$" |
grep -v "^\s*#"
```

8.3　コンピュートノードのネットワーク設定

Nova の設定ファイルの内容を Neutron を利用するように変更します。

```
compute# vi /etc/nova/nova.conf
...
[neutron]
url = http://controller:9696
auth_url = http://controller:35357
auth_type = password
project_domain_name = default
user_domain_name = default
region_name = RegionOne
project_name = service
username = neutron
password = password    ← neutron ユーザーのパスワード
```

次のコマンドで正しく設定を行ったか確認します。

```
compute# less /etc/nova/nova.conf | grep -v "^\s*$" | grep -v "^\s*#"
```

8.4　コンピュートノードの Neutron と関連サービスを再起動

ネットワーク設定を反映させるため、コンピュートノードの Neutron と関連のサービスを再起動します。

```
compute# service nova-compute restart
compute# service neutron-linuxbridge-agent restart
```

8.5　ログの確認

エラーが出ていないかログを確認します。

```
compute# tailf /var/log/nova/nova-compute.log
compute# tailf /var/log/neutron/neutron-linuxbridge-agent.log
```

8.6 Neutronサービスの動作を確認

neutron agent-list コマンドを実行してNeutronエージェントが正しく認識されており、稼働していることを確認します。

```
controller# source admin-openrc.sh
controller# neutron agent-list -c host -c alive -c binary
+------------+-------+---------------------------+
| host       | alive | binary                    |
+------------+-------+---------------------------+
| controller | :-)   | neutron-metadata-agent    |
| controller | :-)   | neutron-linuxbridge-agent |
| controller | :-)   | neutron-dhcp-agent        |
| controller | :-)   | neutron-l3-agent          |
| compute    | :-)   | neutron-linuxbridge-agent |
+------------+-------+---------------------------+
```

※コントローラーとコンピュートで追加され、neutron-linuxbridge-agentが正常に稼働していることが確認できれば問題ありません。念のためログも確認してください。

第9章 仮想ネットワーク設定（コントローラーノード）

　OpenStack Neutron 環境ができたので、OpenStack 内で利用するネットワークを作成します。ネットワークは外部ネットワークと接続するためのパブリックネットワークと、インスタンス間やルーター、内部 DHCP サーバー間の通信に利用するインスタンス用ネットワークの二つを作成します。

　パブリックネットワークは既存のネットワークから一定の範囲のネットワークを OpenStack に割り当てます。ネットワークのゲートウェイ IP アドレス、IP アドレスのセグメントと割り当てる IP アドレスの範囲を決めておく必要があります。例えば 192.168.1.0/24 というネットワークであればゲートウェイ IP アドレスは 192.168.1.1 か 192.168.1.254 がよく使われ、Windows なら ipconfig コマンド、Linux や Unix では ifconfig コマンドで確認できます。パブリックネットワーク用に割り当てる IP アドレスの範囲については、そのネットワークで DHCP サーバーが動いている場合は DHCP サーバーが配る **IP アドレスの範囲を除いた**ネットワークを切り出して利用するようにしてください。

　インスタンスにはインスタンス用ネットワークの範囲の IP アドレスが DHCP サーバーから DHCP Agent を介して割り当てられます。このインスタンスにパブリックネットワークの範囲から Floating IP アドレスを割り当てることで、NAT 接続でインスタンスが外部ネットワークとやり取りができるようになります。

第 9 章　仮想ネットワーク設定（コントローラーノード）

9.1　パブリックネットワークの設定

admin 環境変数ファイルの読み込み

まずは Floating IP アドレス割当用のネットワークである、パブリックネットワークを admin 権限で作成するために admin 環境変数を読み込みます。

```
controller# source admin-openrc.sh
```

パブリックネットワークの作成

ext-net と言う名前でパブリックネットワークを作成します。provider:physical_network で指定する設定は /etc/neutron/plugins/ml2/linuxbridge_agent.ini の physical_interface_mappings に指定した値を設定します。例えば provider:eth0 と設定した場合は provider を指定します。

```
controller(admin)# neutron net-create ext-net --router:external \
 --provider:physical_network provider --provider:network_type flat
Created a new network:
+---------------------------+--------------------------------------+
| Field                     | Value                                |
+---------------------------+--------------------------------------+
| admin_state_up            | True                                 |
| created_at                | 2016-05-18T06:41:05                  |
| description               |                                      |
| id                        | 6d6bee75-55ec-41c4-b738-8409ee051149 |
| ipv4_address_scope        |                                      |
| ipv6_address_scope        |                                      |
| is_default                | False                                |
| mtu                       | 1500                                 |
| name                      | ext-net                              |
| port_security_enabled     | True                                 |
| provider:network_type     | flat                                 |
| provider:physical_network | provider                             |
| provider:segmentation_id  |                                      |
| router:external           | True                                 |
| shared                    | False                                |
| status                    | ACTIVE                               |
| subnets                   |                                      |
| tags                      |                                      |
| tenant_id                 | 360855029dbe44649a5dc862c5a3caf1     |
| updated_at                | 2016-05-18T06:41:05                  |
+---------------------------+--------------------------------------+
```

パブリックネットワーク用サブネットの作成

ext-subnet という名前でパブリックネットワーク用サブネットを作成します。allocation-pool には FloatingIP アドレスとして利用するネットワークの範囲、gateway には指定した範囲のネットワークのゲートウェイ IP アドレスとネットワークセグメントを指定します。

```
controller(admin)# neutron subnet-create ext-net --name ext-subnet \
  --allocation-pool start=10.0.0.200,end=10.0.0.250 \
  --disable-dhcp --gateway 10.0.0.1 10.0.0.0/24

Created a new subnet:
+-------------------+------------------------------------------------+
| Field             | Value                                          |
+-------------------+------------------------------------------------+
| allocation_pools  | {"start": "10.0.0.200", "end": "10.0.0.250"}   |
| cidr              | 10.0.0.0/24                                    |
| created_at        | 2016-05-18T06:51:01                            |
| description       |                                                |
| dns_nameservers   |                                                |
| enable_dhcp       | False                                          |
| gateway_ip        | 10.0.0.1                                       |
| host_routes       |                                                |
| id                | 6c74fe33-47ac-4c28-b396-15b38b1b602f           |
| ip_version        | 4                                              |
| ipv6_address_mode |                                                |
| ipv6_ra_mode      |                                                |
| name              | ext-subnet                                     |
| network_id        | 6d6bee75-55ec-41c4-b738-8409ee051149           |
| subnetpool_id     |                                                |
| tenant_id         | 360855029dbe44649a5dc862c5a3caf1               |
| updated_at        | 2016-05-18T06:51:01                            |
+-------------------+------------------------------------------------+
```

9.2　インスタンス用ネットワークの設定

demo ユーザーと環境変数ファイルの読み込み

次にインスタンス用ネットワークを作成します。インスタンス用ネットワークを作成するために demo 環境変数を読み込みます。

```
controller# source demo-openrc.sh
```

インスタンス用ネットワークの作成

demo-net という名前でインスタンス用ネットワークを作成します。

第 9 章　仮想ネットワーク設定（コントローラーノード）

```
controller(demo)# neutron net-create demo-net
Created a new network:
+---------------------------+--------------------------------------+
| Field                     | Value                                |
+---------------------------+--------------------------------------+
| admin_state_up            | True                                 |
| created_at                | 2016-05-18T06:52:08                  |
| description               |                                      |
| id                        | 8717f502-f07e-4b2f-84fe-b51833db18fb |
| ipv4_address_scope        |                                      |
| ipv6_address_scope        |                                      |
| mtu                       | 1450                                 |
| name                      | demo-net                             |
| port_security_enabled     | True                                 |
| router:external           | False                                |
| shared                    | False                                |
| status                    | ACTIVE                               |
| subnets                   |                                      |
| tags                      |                                      |
| tenant_id                 | 4e00692779864ed185356843ebeb0e2f     |
| updated_at                | 2016-05-18T06:52:08                  |
+---------------------------+--------------------------------------+
```

インスタンス用ネットワークのサブネットを作成

demo-subnet という名前でインスタンス用ネットワークサブネットを作成します。

gateway には指定したインスタンス用ネットワークのサブネットの範囲から任意の IP アドレスを指定します。第 4 オクテットとして 1 を指定した IP アドレスを設定するのが一般的です。ここでは 192.168.0.0/24 のネットワークをインスタンス用ネットワークとして定義し、ゲートウェイ IP として 192.168.0.1 を設定しています。

dns-nameserver には外部ネットワークに接続する場合に名前引きするための DNS サーバーを指定します。ここでは Google Public DNS の 8.8.8.8 を指定していますが、外部の名前解決ができる DNS サーバーであれば何を指定しても構いません。

インスタンス用ネットワーク内で DHCP サーバーが稼働し、インスタンスが起動した時にその DHCP サーバーが dns-nameserver に指定した DNS サーバーと 192.168.0.0/24 の範囲から IP アドレスを確保してインスタンスに割り当てます。

```
controller(demo)# neutron subnet-create demo-net 192.168.0.0/24 \
--name demo-subnet --gateway 192.168.0.1 --dns-nameserver 8.8.8.8
Created a new subnet:
+-------------------+------------------------------------------------+
| Field             | Value                                          |
+-------------------+------------------------------------------------+
| allocation_pools  | {"start": "192.168.0.2", "end": "192.168.0.254"} |
```

```
| cidr             | 192.168.0.0/24                       |
| created_at       | 2016-05-18T06:52:57                  |
| description      |                                      |
| dns_nameservers  | 8.8.8.8                              |
| enable_dhcp      | True                                 |
| gateway_ip       | 192.168.0.1                          |
| host_routes      |                                      |
| id               | f5e3f465-38c7-4915-abc5-6f966412c330 |
| ip_version       | 4                                    |
| name             | demo-subnet                          |
| network_id       | 8717f502-f07e-4b2f-84fe-b51833db18fb |
| subnetpool_id    |                                      |
| tenant_id        | 4e00692779864ed185356843ebeb0e2f     |
| updated_at       | 2016-05-18T06:52:57                  |
```

9.3　仮想ネットワークルーターの設定

　仮想ネットワークルーターを作成して外部接続用ネットワークとインスタンス用ネットワークをルーターに接続し、双方でデータのやり取りを行えるようにします。

demo-router を作成

　仮想ネットワークルーターを作成します。

```
controller(demo)# neutron router-create demo-router
Created a new router:
+-----------------------+--------------------------------------+
| Field                 | Value                                |
+-----------------------+--------------------------------------+
| admin_state_up        | True                                 |
| description           |                                      |
| external_gateway_info |                                      |
| id                    | 92a5c675-94e2-4894-b377-d7ccea194fca |
| name                  | demo-router                          |
| routes                |                                      |
| status                | ACTIVE                               |
| tenant_id             | 4e00692779864ed185356843ebeb0e2f     |
+-----------------------+--------------------------------------+
```

demo-router にサブネットを追加

　仮想ネットワークルーターにインスタンス用ネットワークを接続します。

第 9 章　仮想ネットワーク設定（コントローラーノード）

```
controller(demo)# neutron router-interface-add demo-router demo-subnet
Added interface 5886789d-5093-4c56-a396-2012dba2381c to router demo-router.
```

demo-router にゲートウェイを追加

仮想ネットワークルーターに外部ネットワークを接続します。

```
controller(demo)# neutron router-gateway-set demo-router ext-net
Set gateway for router demo-router
```

9.4　ネットワークの確認

- 仮想ルーターのゲートウェイ IP アドレスの確認

neutron router-port-list コマンドを実行すると、仮想ルーターのそれぞれのポートに割り当てられた IP アドレスを確認することができます。Set gateway for router demo-router コマンドの実行結果から 192.168.0.1 がインスタンスネットワーク側ゲートウェイ IP アドレス、10.0.0.200 がパブリックネットワーク側ゲートウェイ IP アドレスであることがわかります。

作成したネットワークの確認のために、外部 PC からパブリックネットワーク側ゲートウェイ IP アドレスに ping を飛ばしてみましょう。問題なければ仮想ルーターと外部ネットワークとの接続ができていると判断することができます。

```
controller(admin)# source admin-openrc.sh
controller(admin)# neutron router-port-list demo-router -c fixed_ips --max-width 30
+------------------------------+
| fixed_ips                    |
+------------------------------+
| {"subnet_id": "f5e3f465-38c7-4|
| 915-abc5-6f966412c330",      |
| "ip_address": "192.168.0.1"} |
| {"subnet_id": "6c74fe33-47ac-|
| 4c28-b396-15b38b1b602f",     |
| "ip_address": "10.0.0.200"}  |
+------------------------------+

# ping -c3 10.0.0.200|grep "packet loss"
3 packets transmitted, 3 received, 0% packet loss, time 1999ms
(ルーターゲートウェイ宛に各ノードから ping コマンドの実行)
```

※応答が返ってくれば問題ありません。

コントローラーノードで次のようにコマンドを実行すると、従来の Open vSwitch のように linuxbridge の中をのぞくことができます。仮想ルーターと仮想 DHCP サーバーの状態を確認

できます。接続がうまくいかない場合におためしください。

```
controller# ip netns
qrouter-92a5c675-94e2-4894-b377-d7ccea194fca
qdhcp-8717f502-f07e-4b2f-84fe-b51833db18fb

(仮想ルーターと仮想 DHCP サーバーを確認)
...
controller# ip netns exec `ip netns|grep qrouter*` bash
(qrouter にログイン)
...
controller# ip -f inet addr
1: lo: <LOOPBACK,UP,LOWER_UP> mtu 65536 qdisc noqueue state UNKNOWN group default
    inet 127.0.0.1/8 scope host lo
       valid_lft forever preferred_lft forever
2: qr-5886789d-50: <BROADCAST,MULTICAST,UP,LOWER_UP> mtu 1500 qdisc pfifo_fast state UP group default qlen 1000

    inet 192.168.0.1/24 brd 192.168.0.255 scope global qr-5886789d-50

       valid_lft forever preferred_lft forever
3: qg-c39b9d87-8e: <BROADCAST,MULTICAST,UP,LOWER_UP> mtu 1500 qdisc pfifo_fast state UP group default qlen 1000

    inet 10.0.0.200/24 brd 10.0.0.255 scope global qg-c39b9d87-8e

       valid_lft forever preferred_lft forever
(ネットワークデバイスと IP アドレスを表示)
...
controller# ping 8.8.8.8 -I qg-c39b9d87-8e
(Public 側から外にアクセスできることを確認)
```

9.5 インスタンスの起動を確認

　起動イメージ、コンピュート、Neutron ネットワークといった OpenStack の最低限の構成ができあがったので、ここで OpenStack 環境がうまく動作しているか確認しましょう。まずはコマンドを使ってインスタンスを起動するために必要な情報を集める所から始めます。環境設定ファイルを読み込んで、各コマンドを実行し、情報を集めてください。

```
controller# source demo-openrc.sh
controller(demo)# openstack image list
(起動イメージ一覧を表示)
+--------------------------------------+-------------------+--------+
| ID                                   | Name              | Status |
+--------------------------------------+-------------------+--------+
| 12d67808-c7d2-44b4-8bc6-b0ced1878f06 | cirros-0.3.4-x86_64 | active |
+--------------------------------------+-------------------+--------+
```

第 9 章　仮想ネットワーク設定（コントローラーノード）

```
controller(demo)# openstack network list -c ID -c Name
(ネットワーク一覧を表示)
+--------------------------------------+---------+
| ID                                   | Name    |
+--------------------------------------+---------+
| 6d6bee75-55ec-41c4-b738-8409ee051149 | ext-net |
| 8717f502-f07e-4b2f-84fe-b51833db18fb | demo-net|
+--------------------------------------+---------+

controller(demo)# openstack security group list -c ID -c Name
(セキュリティグループ一覧を表示)
+--------------------------------------+---------+
| ID                                   | Name    |
+--------------------------------------+---------+
| 77c03d2e-0898-4e5a-b93a-8bf0c5f26ee2 | default |
+--------------------------------------+---------+

controller(demo)# openstack flavor list -c Name -c Disk
(フレーバー一覧を表示)
+-----------+------+
| Name      | Disk |
+-----------+------+
| m1.tiny   |   1  |
| m1.small  |  20  |
| m1.medium |  40  |
| m1.large  |  80  |
| m1.xlarge | 160  |
+-----------+------+
```

nova boot コマンドを使って、インスタンスを起動します。正常に起動したら nova delete コマンドでインスタンスを削除してください。

```
controller(demo)# nova boot --flavor m1.tiny --image "cirros-0.3.4-x86_64" \
 --nic net-id=8717f502-f07e-4b2f-84fe-b51833db18fb \
 --security-group 77c03d2e-0898-4e5a-b93a-8bf0c5f26ee2 vm1

(インスタンスを起動)

controller(demo)# watch nova list
(インスタンス一覧を表示)
+--------------------------------------+------+--------+------------+-------------+-
| ID                                   | Name | Status | Task State | Power State |
Networks                               |
+--------------------------------------+------+--------+------------+-------------+-
| 2bc42415-c6a7-45a2-b5ee-b865a26bf101 | vm1  | ACTIVE | -          | Running     |
demo-net=192.168.0.3 |
+--------------------------------------+------+--------+------------+-------------+-

# grep "ERROR\|WARNING" /var/log/rabbitmq/*.log
# grep "ERROR\|WARNING" /var/log/neutron/*
# grep "ERROR\|WARNING" /var/log/nova/*
(各ノードの関連サービスでエラーが出ていないことを確認)
```

```
controller(demo)# nova delete vm1
Request to delete server vm1 has been accepted.
(起動したインスタンスを削除)
```

第10章 Cinder のインストール（コントローラーノード）

10.1 データベースを作成

MariaDB のデータベースに Cinder のデータベースを作成します。

```
controller# mysql -u root -p << EOF
CREATE DATABASE cinder;
GRANT ALL PRIVILEGES ON cinder.* TO 'cinder'@'localhost' \
  IDENTIFIED BY 'password';
GRANT ALL PRIVILEGES ON cinder.* TO 'cinder'@'%' \
  IDENTIFIED BY 'password';
EOF
Enter password: ← MariaDB の root パスワード password を入力
```

10.2 データベースの確認

MariaDB に Cinder のデータベースが登録されたか確認します。

```
controller# mysql -u cinder -p
Enter password: ← MariaDB の cinder パスワード password を入力
...
Type 'help;' or '\h' for help. Type '\c' to clear the current input statement.

MariaDB [(none)]> show databases;
+--------------------+
| Database           |
+--------------------+
| information_schema |
| cinder             |
+--------------------+
```

```
2 rows in set (0.00 sec)
```

※ユーザー cinder でログイン可能でデータベースの閲覧が可能なら問題ありません。

10.3　Cinder サービスなどの作成

- admin 環境変数の読み込み

```
controller# source admin-openrc.sh
```

- cinder ユーザーの作成

```
controller# openstack user create --domain default --password-prompt cinder
User Password: password    ← cinder ユーザーのパスワードを設定 (本書は password を設定)
Repeat User Password: password
+-----------+----------------------------------+
| Field     | Value                            |
+-----------+----------------------------------+
| domain_id | c12fc3119bb046a09ba53e0566004f76 |
| enabled   | True                             |
| id        | 4b798308a76041d1a0dc86464ac1208a |
| name      | cinder                           |
+-----------+----------------------------------+
```

- admin ロールを cinder ユーザーに追加

```
controller# openstack role add --project service --user cinder admin
```

- cinder および cinder v2 サービスの作成

```
controller# openstack service create --name cinder \
--description "OpenStack Block Storage" volume
+-------------+----------------------------------+
| Field       | Value                            |
+-------------+----------------------------------+
| description | OpenStack Block Storage          |
| enabled     | True                             |
| id          | f04a5af43a774b33887e4e42a2478340 |
| name        | cinder                           |
| type        | volume                           |
+-------------+----------------------------------+
```

```
controller# openstack service create --name cinderv2 \
--description "OpenStack Block Storage" volumev2
+-------------+----------------------------------+
| Field       | Value                            |
+-------------+----------------------------------+
| description | OpenStack Block Storage          |
| enabled     | True                             |
| id          | be8bebb2f37345b3a1f14512193d86f6 |
| name        | cinderv2                         |
| type        | volumev2                         |
+-------------+----------------------------------+
```

- Block Storage サービスの API エンドポイントを作成

```
controller# openstack endpoint create --region RegionOne \
 volume public http://controller:8776/v1/%\(tenant_id\)s
controller# openstack endpoint create --region RegionOne \
 volume internal http://controller:8776/v1/%\(tenant_id\)s
controller# openstack endpoint create --region RegionOne \
 volume admin http://controller:8776/v1/%\(tenant_id\)s
```

```
controller# openstack endpoint create --region RegionOne \
   volumev2 public http://controller:8776/v2/%\(tenant_id\)s
controller# openstack endpoint create --region RegionOne \
   volumev2 internal http://controller:8776/v2/%\(tenant_id\)s
controller# openstack endpoint create --region RegionOne \
   volumev2 admin http://controller:8776/v2/%\(tenant_id\)s
```

10.4　パッケージのインストール

　本書では Block Storage コントローラーと Block Storage ボリュームコンポーネントを一台のマシンで構築するため、両方の役割をインストールします。

```
controller# apt-get install lvm2 cinder-api cinder-scheduler cinder-volume
python-mysqldb python-cinderclient
```

10.5　Cinderの設定を変更

```
controller# vi /etc/cinder/cinder.conf

[DEFAULT]
...
auth_strategy = keystone          ←確認
```

```
#lock_path = /var/lock/cinder    ←コメントアウト

↓↓ 以下追記 ↓↓

rpc_backend = rabbit

my_ip = 10.0.0.101      #コントローラーノード
enabled_backends = lvm

[oslo_messaging_rabbit]
rabbit_host = controller
rabbit_userid = openstack
rabbit_password = password

[oslo_concurrency]
lock_path = /var/lib/cinder/tmp

[database]
connection = mysql+pymysql://cinder:password@controller/cinder

[keystone_authtoken]
auth_uri = http://controller:5000
auth_url = http://controller:35357
memcached_servers = controller:11211
auth_type = password
project_domain_name = default
user_domain_name = default
project_name = service
username = cinder
password = password          ← cinder ユーザーのパスワード

[lvm]
volume_driver = cinder.volume.drivers.lvm.LVMVolumeDriver
volume_group = cinder-volumes
iscsi_protocol = iscsi
iscsi_helper = tgtadm
```

次のコマンドで正しく設定を行ったか確認します。

```
controller# less /etc/cinder/cinder.conf | grep -v "^\s*$" | grep -v "^\s*#"
```

10.6　データベースに展開

```
controller# su -s /bin/sh -c "cinder-manage db sync" cinder
```

10.7 使用しないデータベースファイルを削除

```
controller# rm /var/lib/cinder/cinder.sqlite
```

10.8 nova-api の設定変更

controller ノードの /etc/nova/nova.conf に以下を追記します。

```
controller# vi /etc/nova/nova.conf

(以下追記)

[cinder]
os_region_name = RegionOne
```

10.9 サービスの再起動

設定を反映させるために、Cinder のサービスを再起動します。

```
controller# service cinder-scheduler restart && service cinder-api restart && service nova-api restart
```

10.10 イメージ格納用ボリュームの作成

　イメージ格納用ボリュームを設定するために物理ボリュームの設定、ボリュームの作成を行います。

物理ボリュームを追加

　本書ではコントローラーノードにハードディスクを追加して、そのボリュームを Cinder 用ボリュームとして使います。コントローラーノードを一旦シャットダウンしてからハードディスクを増設し、再起動してください。新しい増設したディスクは dmesg コマンドなどを使って確認できます。

第10章 Cinderのインストール（コントローラーノード）

```
controller# dmesg |grep sd|grep "logical blocks"
[    1.361779] sd 2:0:0:0: [sda] 62914560 512-byte logical blocks: (32.2 GB/30.0
GiB)  ←システムディスク
[    1.362105] sd 2:0:1:0: [sdb] 33554432 512-byte logical blocks: (17.1 GB/16.0
GiB)  ←追加ディスク
```

仮想マシンにハードディスクを増設した場合は/dev/vdbなどのようにデバイス名が異なる場合があります。

物理ボリュームを設定

以下コマンドで物理ボリュームを作成します。

- LVM物理ボリュームの作成

```
controller# pvcreate /dev/sdb
  Physical volume "/dev/sdb" successfully created
```

- LVMボリュームグループの作成

```
controller# vgcreate cinder-volumes /dev/sdb
  Volume group "cinder-volumes" successfully created
```

- /etc/lvm/lvm.confにデバイスを指定

```
devices {
...
filter = [ "a/sdb/", "r/.*/" ]
```

Cinder-Volumeサービスの再起動

Cinderストレージの設定を反映させるために、Cinder-Volumeのサービスを再起動します。

```
controller# service cinder-volume restart && service tgt restart
```

admin環境変数設定ファイルを読み込み

adminのみ実行可能なコマンドを実行するために、admin環境変数を読み込みます。

10.10 イメージ格納用ボリュームの作成

```
controller# source admin-openrc.sh
```

Cinder サービスの確認

以下コマンドで Cinder サービスの一覧を表示し、正常に動作していることを確認します。

```
controller# cinder service-list
+------------------+----------------+------+---------+-------+----------------------+
|      Binary      |      Host      | Zone | Status  | State |     Updated_at       |
+------------------+----------------+------+---------+-------+----------------------+
| cinder-scheduler |   controller   | nova | enabled |  up   |  2016-05-19T02:39    |
|  cinder-volume   | controller@lvm | nova | enabled |  up   |  2016-05-19T02:39    |
+------------------+----------------+------+---------+-------+----------------------+
```

Cinder ボリュームの作成

以下コマンドで Cinder ボリュームを作成し、正常に Cinder が動作していることを確認します。

```
controller# openstack volume create --size 1 volume
(1GB のストレージを作成)
+---------------------+--------------------------------------+
| Field               | Value                                |
+---------------------+--------------------------------------+
| attachments         | []                                   |
| availability_zone   | nova                                 |
| bootable            | false                                |
| consistencygroup_id | None                                 |
| created_at          | 2016-05-27T06:26:22.501800           |
| description         | None                                 |
| encrypted           | False                                |
| id                  | cb0979fa-ea2e-4c52-a26a-c815c53a22ee |
| migration_status    | None                                 |
| multiattach         | False                                |
| name                | volume                               |
| properties          |                                      |
| replication_status  | disabled                             |
| size                | 1                                    |  ← 作成できた
| snapshot_id         | None                                 |
| source_volid        | None                                 |
| status              | creating                             |
| type                | None                                 |
| updated_at          | None                                 |
| user_id             | 197e779ea83e43d98197b5abe23de80c     |
+---------------------+--------------------------------------+
```

第11章 Dashboard のインストールと確認（コントローラーノード）

クライアントマシンからブラウザーで OpenStack 環境を操作可能な Web インターフェイスをインストールします。

11.1 パッケージのインストール

コントローラーノードに Dashboard をインストールします。

```
controller# apt-get install openstack-dashboard
```

11.2 Dashboardの設定を変更

インストールした Dashboard の設定を変更します。

```
controller# vi /etc/openstack-dashboard/local_settings.py

(以下変更)

...

OPENSTACK_HOST = "controller"          ← 変更
OPENSTACK_KEYSTONE_URL = "http://%s:5000/v3" % OPENSTACK_HOST   ← 変更
ALLOWED_HOSTS = ['*', ]

SESSION_ENGINE = 'django.contrib.sessions.backends.cache'   ←追記
CACHES = {                                        ← CACHES パラメーターのコメントを外す
    'default': {
        'BACKEND': 'django.core.cache.backends.memcached.MemcachedCache',
```

第 11 章　Dashboard のインストールと確認（コントローラーノード）

```
        'LOCATION': 'controller:11211',  ← 変更
    },
}
OPENSTACK_API_VERSIONS = {         ← OPENSTACK_API_VERSIONS パラメーターのコメントを外す
    "identity": 3,   ← 変更
    "volume": 2,     ← 変更
    "compute": 2,    ← 変更
}

OPENSTACK_KEYSTONE_DEFAULT_DOMAIN = 'default'   ← コメントを外す
OPENSTACK_KEYSTONE_DEFAULT_ROLE = "user"   ← 変更
TIME_ZONE = "Asia/Tokyo"   ← 変更
```

次のコマンドで正しく設定を行ったか確認します。

```
controller# less /etc/openstack-dashboard/local_settings.py  | grep -v "^\s*$" | grep -v "^\s*#"
```

念のため、リダイレクトするように設定しておきます（数字は待ち時間）。

```
controller# vi /var/www/html/index.html
...
  <head>
    <meta http-equiv="Content-Type" content="text/html; charset=UTF-8" />
    <meta http-equiv="refresh" content="3; url=/horizon" />    ← 追記
```

設定変更を反映させるため、Apache Web サーバーを再起動します。

```
controller# service apache2 restart
```

11.3　Dashboardにアクセス

コントローラーノードとネットワークに接続されているマシンからブラウザで以下 URL に接続して OpenStack のログイン画面が表示されるか確認します。

※ブラウザで接続するマシンは予め DNS もしくは/etc/hosts にコントローラーノードの IP を記述しておく等コンピュートノードの名前解決を行っておく必要があります。

```
http://controller/horizon/
```

※上記 URL にアクセスしてログイン画面が表示され、ユーザー admin と demo でログイン（パスワード:password）でログインできれば問題ありません。

11.4 セキュリティグループの設定

OpenStack で動作するインスタンスのファイアウォール設定は、セキュリティグループで行います。ログイン後、次の手順でセキュリティグループを設定できます。

1. demo ユーザーでログイン
2. 「プロジェクト→コンピュート→アクセスとセキュリティ」を選択
3. 「ルールの管理」ボタンをクリック
4. 「ルールの追加」で許可するルールを定義
5. 「追加」ボタンをクリック

インスタンスに対して Ping を実行したい場合はルールとしてすべての ICMP を、インスタンスに SSH 接続したい場合は SSH をルールとしてセキュリティグループに追加してください。

セキュリティーグループは複数作成できます。作成したセキュリティーグループをインスタンス起動時に選択することで、セキュリティグループで定義したポートを解放したり、拒否したり、接続できるクライアントを制限することができます。

11.5 キーペアの作成

OpenStack ではインスタンスへのアクセスはデフォルトで公開鍵認証方式で行います。次の手順でキーペアを作成できます。

1. demo ユーザーでログイン
2. 「プロジェクト→コンピュート→アクセスとセキュリティ」をクリック
3. 「キーペア」タブをクリック
4. 「キーペアの作成」ボタンをクリック
5. キーペア名を入力
6. 「キーペアの作成」ボタンをクリック
7. キーペア（拡張子:pem）ファイルをダウンロード

11.6 インスタンスの起動

前の手順（4.9 イメージの取得と登録）で Glance に CirrOS イメージを登録していますので、

第 11 章　Dashboard のインストールと確認（コントローラーノード）

早速構築した OpenStack 環境上でインスタンスを起動してみましょう。

1. demo ユーザーでログイン
2. 「プロジェクト→コンピュート→イメージ」をクリック
3. イメージ一覧から起動する OS イメージを選び、「インスタンスの起動」ボタンをクリック
4. 「インスタンスの起動」詳細タブで起動するインスタンス名、フレーバー、インスタンス数を設定
5. アクセスとセキュリティタブで割り当てるキーペア、セキュリティーグループを設定
6. ネットワークタブで割り当てるネットワークを設定
7. 作成後タブで必要に応じてユーザーデータの入力（オプション）
8. 高度な設定タブでパーティションなどの構成を設定（オプション）
9. 右下の「起動」ボタンをクリック

11.7　Floating IP の設定

起動したインスタンスに Floating IP アドレスを設定することで、Dashboard のコンソール以外からインスタンスにアクセスできるようになります。インスタンスに Floating IP を割り当てるには次の手順で行います。

1. demo ユーザーでログイン
2. 「プロジェクト→コンピュート→インスタンス」をクリック
3. インスタンスの一覧から割り当てるインスタンスをクリック
4. アクションメニューから「Floating IP の割り当て」をクリック
5. 「Floating IP 割り当ての管理」画面の IP アドレスで「+」ボタンをクリック
6. 右下の「IP の確保」ボタンをクリック
7. 割り当てる IP アドレスとインスタンスを選択して右下の「割り当て」ボタンをクリック

11.8　インスタンスへのアクセス

Floating IP を割り当てて、かつセキュリティグループの設定を適切に行っていれば、リモートアクセスできるようになります。セキュリティーグループで SSH を許可した場合、端末から SSH 接続が可能になります（下記は実行例）。

```
client$ ssh -i mykey.pem cloud-user@instance-floating-ip
```

その他、適切なポートを開放してインスタンスへの Ping を許可したり、インスタンスで Web サーバーを起動して外部 PC からアクセスしてみましょう。

11.9　Ubuntu テーマの削除

Ubuntu Server ベースで OpenStack 環境を構築した場合、OpenStack Dashboard は Ubuntu のテーマが設定されています。このテーマは OpenStack 標準テーマと比べてフォントが大きく見やすくて良いのですが、一部のボタンなどでボタンラベルが表示されない問題が発生することがあります。

図 11.1　Ubuntu テーマの削除

次のように実行すると Ubuntu のカスタムテーマを削除できます。必要に応じて対応してください。

```
$ sudo apt-get remove --auto-remove openstack-dashboard-ubuntu-theme
```

第II部

監視環境 構築編

第12章 Zabbixのインストール

ZabbixはZabbix SIA社が提供するパッケージを使う方法とCanonical Ubuntuが提供するパッケージを使う方法がありますが、今回は前者を利用してZabbixが動作する環境を作っていきましょう。

本例ではZabbixをUbuntu Server 14.04.4上にオールインワン構成でセットアップする手順を示します。

12.1 パッケージのインストール

次のコマンドを実行し、Zabbix3.0のパッケージをインストールします。

```
zabbix# wget
http://repo.zabbix.com/zabbix/3.0/ubuntu/pool/main/z/zabbix-release/zabbix-release_3.0-1+trusty_all.deb
zabbix# dpkg -i zabbix-release_3.0-1+trusty_all.deb
```

次に、Zabbixの動作に必要となるパッケージ群をインストールします。

MySQLインストール中にパスワードの入力を要求されますので、MySQLのrootユーザーに対するパスワードを設定します。本書ではパスワードとして「password」を設定します。

```
zabbix# apt-get update
zabbix# apt-get install php5-mysql zabbix-agent zabbix-server-mysql \
 zabbix-java-gateway zabbix-frontend-php
```

12.2 Zabbix用データベースの作成

データベースの作成

次のコマンドを実行し、Zabbix用MySQLユーザおよびデータベースを作成します。

```
zabbix# mysql -u root -p << EOF
CREATE DATABASE zabbix;
GRANT ALL PRIVILEGES ON zabbix.* TO 'zabbix'@'localhost' \
  IDENTIFIED BY 'zabbix';
EOF
Enter password: ← MySQLのrootパスワードを入力 (12-1で設定したもの)
```

次のコマンドを実行し、Zabbix用データベースにテーブル等のデータベースオブジェクトを作成します。

```
# zcat /usr/share/doc/zabbix-server-mysql/create.sql.gz | mysql -uroot zabbix -p
Enter password:← パスワードzabbixを入力
```

データベースの確認

作成したデータベーステーブルにアクセスしてみましょう。zabbixデータベースの各テーブルが全て参照可能であれば問題ありません。

```
zabbix# mysql -u root -p
Enter password:              ← パスワードzabbixを入力
mysql> show databases;
+--------------------+
| Database           |
+--------------------+
| information_schema |
| mysql              |
| performance_schema |
| zabbix             |
+--------------------+
4 rows in set (0.00 sec)
mysql> use zabbix;
mysql> show tables;
+----------------------+
| Tables_in_zabbix     |
+----------------------+
| acknowledges         |
| actions              |
| alerts               |
...

mysql> describe acknowledges;
+---------------+---------------------+------+-----+---------+-------+
```

```
| Field          | Type             | Null | Key | Default | Extra |
+----------------+------------------+------+-----+---------+-------+
| acknowledgeid  | bigint(20) unsigned | NO   | PRI | NULL    |       |
| userid         | bigint(20) unsigned | NO   | MUL | NULL    |       |
| eventid        | bigint(20) unsigned | NO   | MUL | NULL    |       |
| clock          | int(11)          | NO   | MUL | 0       |       |
| message        | varchar(255)     | NO   |     |         |       |
+----------------+------------------+------+-----+---------+-------+
5 rows in set (0.00 sec)
```

12.3　Zabbixサーバーの設定および起動

/etc/zabbix/zabbix_server.conf を編集し、次の行を追加します。なお、MySQL ユーザ zabbix のパスワードを別の文字列に変更した場合は、該当文字列を指定する必要があります。

```
zabbix# vi /etc/zabbix/zabbix_server.conf
...
DBPassword=zabbix
```

これまでの設定変更を反映させるため、サービス zabbix-server を再起動します。

```
zabbix# service zabbix-server restart
```

12.4　Zabbix frontendの設定および起動

Zabbix の動作要件を満たすため、/etc/apache2/conf-enabled/zabbix.conf を編集しPHP の設定を変更します。

```
zabbix# vi /etc/apache2/conf-enabled/zabbix.conf

php_value max_execution_time 300
php_value memory_limit 128M
php_value post_max_size 16M
php_value upload_max_filesize 2M
php_value max_input_time 300
php_value always_populate_raw_post_data -1
# php_value date.timezone Europe/Riga
php_value date.timezone Asia/Tokyo
```

これまでの設定変更を反映させるため、Apache を再起動します。

第 12 章 Zabbix のインストール

```
zabbix# service apache2 restart
```

Web ブラウザで Zabbix frontend へアクセスします。画面指示に従い、Zabbix の初期設定を行います。

```
http://<Zabbix frontend の IP アドレス>/zabbix/
```

Zabbix の初期設定では次のような画面が表示されます。「Next step」ボタンをクリックして次に進みます。

図 12.1 Zabbix 初期セットアップ

- 「2. Check of pre-requisites」は、システム要件を満たしている（全て OK となっている）ことを確認します。
- 「3. Configure DB connection」は次のように設定します。

項目	設定値
Database type	MySQL
Database host	localhost
Database Port	0
Database name	zabbix
User	zabbix
Password	zabbix

- 「4. Zabbix server details」は Zabbix Server のインストール場所の指定です。本例ではそのまま次に進みます。
- 「5. Pre-Installation summary」で設定を確認し、問題なければ次に進みます。
- 「6. Install」で設定ファイルのパスが表示されるので確認し「Finish」ボタンをクリックします（/etc/zabbix/zabbix.conf.php）。
- ログイン画面が表示されるので、Admin/zabbix（初期パスワード）でログインします。

12.5　Zabbix frontend の日本語化設定

図 12.2　言語設定の変更

　Zabbix frontend へログインし、言語設定を日本語にします。画面右上のユーザーアイコンを選択し、「User」→「Language」→「Japanese(ja_JP)」を選択し、Update ボタンを押します。

　言語として「Japanese(ja_JP)」が選択できない場合は、次のように実行してから再度上記の操作を実行してください。

```
zabbix# localedef -f UTF-8 -i ja_JP ja_JP
zabbix# service zabbix-server restart
zabbix# service apache2 restart
```

第13章 Hatoholのインストール

Hatohol 16.04 は CentOS 7 以降、Ubuntu Server 14.04 などで動作します。CentOS 7 向けには公式の RPM パッケージが公開されており、yum コマンドを使ってインストール可能です。本例では Hatohol を CentOS 7 上にオールインワン構成でセットアップする手順を示します。

インストールには次のリポジトリーへのアクセスができる必要があります。

- Hatohol（http://project-hatohol.github.io/repo/）
- EPEL（http://dl.fedoraproject.org/pub/epel/7/x86_64/）
- rabbitmq.com/releases（https://www.rabbitmq.com/releases/rabbitmq-server/）

13.1 インストール

1. Hatohol をインストールするため、まずは wget をインストールし Project Hatohol 公式の yum リポジトリーを登録します。

現時点の Hatohol はバージョンのアップグレードに対応していません。そこで yum update コマンドで想定しない Hatohol のアップグレードが行われないようにするために、リポジトリーを enabled=0 に設定します。

```
hatohol# yum install wget
hatohol# wget -P /etc/yum.repos.d/
http://project-hatohol.github.io/repo/hatohol-el7.repo
hatohol# sed -i s/enabled=1/enabled=0/g /etc/yum.repos.d/hatohol-el7.repo
```

2. EPEL リポジトリー上のパッケージのインストールをサポートするため、EPEL パッ

第13章 Hatohol のインストール

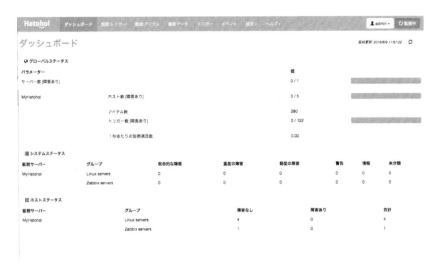

図 13.1　Hatohol ダッシュボード

ケージを追加インストールします。

```
hatohol# yum install epel-release
hatohol# yum update
```

3. Hatohol サーバをインストールします。

```
hatohol# yum --enablerepo=hatohol install hatohol-server
```

4. Hatohol Web Frontend をインストールします。

```
hatohol# yum --enablerepo=hatohol install hatohol-web
```

5. 必要となる追加パッケージをインストールします。

```
hatohol# yum install mariadb-server
```

13.2　MariaDB サーバーの設定

まずローカルの MariaDB 関連の設定を行います。

1. MariaDB の起動

```
hatohol# systemctl enable mariadb
hatohol# systemctl start mariadb
```

2. root ユーザーのパスワードを設定

13.2 MariaDB サーバーの設定

インストール直後は root ユーザーのパスワードは設定されないため、次のコマンドを使って root パスワードの設定を行います。この後の設定は適宜実施します。

```
hatohol# mysql_secure_installation
...
Enter current password for root (enter for none):  ← Enter キーを入力
OK, successfully used password, moving on...

Setting the root password ensures that nobody can log into the MariaDB
root user without the proper authorisation.

Set root password? [Y/n] y
New password:                           ←パスワードを 2 回入力
Re-enter new password:
Password updated successfully!
Reloading privilege tables..
 ... Success!

Remove anonymous users? [Y/n] y
Disallow root login remotely? [Y/n] n
Remove test database and access to it? [Y/n] y
Reload privilege tables now? [Y/n] y
```

3. Hatohol DB の初期化

```
hatohol# hatohol-db-initiator --db_user root --db_password <MariaDB の root パスワード>
```

そのまま上記コマンドを実行した場合、MySQL ユーザ hatohol、データベース hatohol が作成されます。これらを変更する場合、事前に /etc/hatohol/hatohol.conf を編集してください。

4. Hatohol Web 用 DB の作成

```
hatohol# mysql -u root -p
Enter password:     ←設定した root パスワードを入力
MariaDB > CREATE DATABASE hatohol_client DEFAULT CHARACTER SET utf8;
MariaDB > GRANT ALL PRIVILEGES ON hatohol_client.* TO hatohol@localhost IDENTIFIED
BY 'hatohol';
MariaDB > quit;
```

5. Hatohol Web 用 DB へのテーブル追加

```
hatohol# /usr/libexec/hatohol/client/manage.py syncdb
```

6. Hatohol サーバーの自動起動の有効化と起動

```
hatohol# systemctl enable hatohol
hatohol# systemctl start hatohol
```

7. Hatohol Web の自動起動の有効化と起動

```
hatohol# systemctl enable httpd
hatohol# systemctl start httpd
```

確認のため、hatohol のプロセスが起動しているか以下のコマンドを実行します。

```
# systemctl status hatohol | egrep "Active|Main PID"
```

13.3 セキュリティ設定の変更

　CentOS インストール後の初期状態では、SELinux, Firewalld, iptables といったセキュリティ機構により他のコンピュータからのアクセスに制限が加えられます。Hatohol は現時点で SELinux による強制アクセス制御機能が有効化されていると動作しないため、これを解除する必要があります。

SELinux の設定

現在の SELinux の実行モードを確認するには getenforce モードを実行します。

```
hatohol# getenforce
Enforcing
```

SELinux ポリシールールの適用を無効化するには、/etc/selinux/config を編集します。

- 編集前

```
SELINUX=enforcing
```

- 編集後

```
SELINUX=disabled
```

パケットフィルタリングの設定

　フィルタリングの設定変更は、次のコマンドで恒久的に変更可能です。5672 番ポートについては後述の RabbitMQ サービスで使用します。

```
hatohol# firewall-cmd --zone=public --add-port=80/tcp --permanent
hatohol# firewall-cmd --zone=public --add-port=80/tcp
hatohol# firewall-cmd --zone=public --add-port=5672/tcp --permanent
hatohol# firewall-cmd --zone=public --add-port=5672/tcp
```

設定を反映させるため、Hatoholサーバーの再起動をします。

```
hatohol# shutdown -r now
```

13.4 Hatohol Arm Plugin Interface 2 (HAPI2)の設定

RabbitMQのインストール

EPELリポジトリより、RabbitMQとErlangをインストールします。RabbitMQは本書の執筆時点(2016年6月現在)で最新のバージョン3.6.2をインストールします。

- https://www.rabbitmq.com/install-rpm.html

```
hatohol# yum install erlang
hatohol# rpm --import https://www.rabbitmq.com/rabbitmq-signing-key-public.asc
hatohol# yum install
https://www.rabbitmq.com/releases/rabbitmq-server/v3.6.2/rabbitmq-server-3.6.2-1
.noarch.rpm
```

RabbitMQサービスの起動

RabbitMQを有効化します。

```
hatohol# systemctl enable rabbitmq-server
hatohol# systemctl start rabbitmq-server
```

RabbitMQの各種設定

RabbitMQにアクセスするためのユーザーとしてhatoholユーザーを作成し、必要なパーミッションを設定します。以下はRabbitMQのパスワードをhatoholにする例です。

```
hatohol# rabbitmqctl add_vhost hatohol
hatohol# rabbitmqctl add_user hatohol hatohol
hatohol# rabbitmqctl set_permissions -p hatohol hatohol ".*" ".*" ".*"
```

Zabbix プラグインのインストール

以下のコマンドを実行して、Hatohol に HAPI2 の Zabbix プラグインをインストールします。

```
hatohol# yum --enablerepo=hatohol install hatohol-hap2-zabbix
hatohol# systemctl restart hatohol
```

HAPI2 の追加

以下のコマンドを実行して、Hatohol に HAPI2 を追加します。

```
hatohol# hatohol-db-initiator --db_user root --db_password <MariaDB の root パスワード>
hatohol# systemctl restart hatohol
```

13.5　Hatohol による監視情報の閲覧

　Hatohol Web が動作しているホストのトップディレクトリーを Web ブラウザで表示します。10.0.0.10 で動作している場合は、次の URL となります。admin/hatohol（初期パスワード）でログインできます。

```
http://10.0.0.10/hatohol/
```

　Hatohol は監視サーバーから取得したログ、イベント、性能情報を表示するだけでなく、それらの情報を統計してグラフとして出力することができる機能が備わっています。CPU のシステム時間、ユーザー時間をグラフとして出力すると次のようになります。

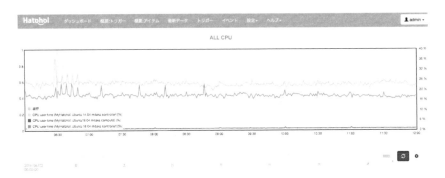

図 13.2　Hatohol のグラフ機能

13.6 HatoholにZabbixサーバーを登録

HatoholをインストールできたらZabbixサーバーの情報を追加します。Hatohol Webにログインしたら、上部のメニューバーの「設定→監視サーバー」をクリックします。「監視サーバー」の画面に切り替わったら「監視サーバー追加」ボタンをクリックしてノードを登録します。

項目	設定値
監視サーバータイプ	Zabbix (HAPI2)
ニックネーム	zabbix
Zabbix API URL	http://(zabbixサーバーのIPアドレス)/zabbix/api_jsonrpc.php
ユーザー名	初期設定:Admin
パスワード	初期設定:zabbix
ポーリング間隔（秒）	30
リトライ間隔（秒）	10
パッシブモード	オフ（チェックを入れない）
ブローカーURL	amqp://hatohol:hatohol@localhost/hatohol
以降空白	

設定が終わったら適用ボタンを押します。

ページを再読み込みして、通信状態が「初期状態」から「正常」になる事を確認します。

図13.3 Zabbixサーバーの追加

13.7 HatoholでZabbixサーバーの監視

インストール直後のZabbixサーバーはモニタリング設定が無効化されています。これを有効化するとZabbixサーバー自身の監視データを取得する事ができるようになり、HatoholでZabbixサーバーの状態を閲覧できるようになります。

Zabbixサーバーのモニタリング設定を変更するには、次の手順で行います。

- Zabbix のメインメニュー「設定 → ホスト」をクリックします。
- ホストグループ一覧から「Zabbix server」をクリックします。
- 「Zabbix server」のホスト設定で、「有効」にチェックボックスを入れます。
- 更新ボタンをクリックして設定変更を適用します。

以上の手順で、Zabbix サーバーを監視対象として設定できます。

13.8　Hatohol でその他のホストの監視

Zabbix と Hatohol の連携ができたので、あとは対象のサーバーに Zabbix Agent をインストールし、手動で Zabbix サーバーにホストを追加するか、ディスカバリ自動登録を使って、特定のネットワークセグメントに所属する Zabbix Agent がインストールされたホストを自動登録するようにセットアップするなどの方法で監視ノードを追加できます。追加したノードは Zabbix および Hatohol で監視する事ができます。

Zabbix Agent のインストール

Zabbix で OpenStack の各ノードを監視するために Zabbix Agent をインストールします。今回は Zabbix 3.0.x をインストールしたので、Zabbix Agent 3.0.x パッケージをインストールします。

```
agent# wget
http://repo.zabbix.com/zabbix/3.0/ubuntu/pool/main/z/zabbix-release/zabbix-release_3.0-1+trusty_all.deb
agent# dpkg -i zabbix-release_3.0-1+trusty_all.deb
agent# apt-get update && apt-get install zabbix-agent
```

Zabbix Agent の設定

Zabbix Agent をインストールしたら次にどの Zabbix サーバーと通信するのか設定を行う必要があります。最低限必要な設定は次の 4 つです。次のように設定します。

(controller ノードの設定記述例)

```
agent# vi /etc/zabbix/zabbix_agentd.conf
...
Server          10.0.0.10       ← Zabbix サーバーの IP アドレスに書き換え
ServerActive    10.0.0.10       ← Zabbix サーバーの IP アドレスに書き換え
Hostname        controller      ← Zabbix サーバーに登録する際のホスト名と同一のものを設定
ListenIP        10.0.0.101      ← Zabbix エージェントが待ち受ける側の IP アドレス
```

13.8 Hatohol でその他のホストの監視

ListenIP には Zabbix サーバーと通信可能な NIC の IP アドレスを指定します。

変更した Zabbix Agent の設定を反映させるため、Zabbix Agent サービスを再起動します。

```
agent# service zabbix-agent restart
```

ホストの登録

Zabbix Agent のセットアップが終わったら、次に Zabbix Agent をセットアップしたサーバーを Zabbix の管理対象として追加します。次のように設定します。

- Zabbix の Web 管理コンソールにアクセスします。
- 「設定」→「ホスト」をクリックします。初期設定時は Zabbix server のみが登録されています。同様に監視対象のサーバーを Zabbix に登録します。
- 画面の右上にある、「ホストの作成」ボタンをクリックします。
- 次のように設定します。

ホストの設定	説明
ホスト名	zabbix_agentd.conf にそれぞれ記述した Hostname を記述
表示名	表示名（オプション）
グループ	所属グループの指定。例として Linux servers を指定
エージェントのインターフェース	監視対象とする Agent がインストールされたホストの IP アドレス（もしくはホスト名）

その他の項目は適宜設定し、「有効」ボタンのチェックボックスをオンにして追加ボタンを押します。

- 「ホスト」の「テンプレート」タブをクリックして設定を切り替えます。
- 「テンプレートとのリンク」の検索ボックスに「Template OS Linux」と入力し、選択肢が出てきたらクリックします。そのほかのテンプレートを割り当てるにはテンプレートを検索し、該当のものを選択します。
- 「テンプレートとのリンク」にテンプレートを追加したら、その項目の「追加」リンクをクリックします。「テンプレートとのリンク」に追加されます。
- 更新ボタンをクリックします。
- ホスト登録画面にサーバーが追加されます。ページの再読み込みを実行して、Zabbix エージェントが有効になっていることを確認します。Zabbbix エージェントの設定が有効になっている場合は「ZBX」アイコンが緑色に変化します。

第 13 章 Hatohol のインストール

図 13.4　Zabbix エージェントステータスを確認

- ほかに追加したいサーバーがあれば「Zabbix Agent のインストール、設定、ホストの登録」の一連の流れを繰り返します。監視したい対象が大量にある場合はオートディスカバリを検討してください。

Hatohol で確認

登録したサーバーの情報が Hatohol で閲覧できるか確認してみましょう。Zabbix サーバー以外のログなど表示できるようになれば正常に閲覧できています。

図 13.5　OpenStack ノードの監視

参考情報

ホストの追加やディスカバリ自動登録については次のドキュメントをご覧ください。

- http://www.zabbix.com/jp/auto_discovery.php
- https://www.zabbix.com/documentation/3.0/manual/quickstart/host
- https://www.zabbix.com/documentation/3.0/manual/discovery/auto_registration
- https://www.zabbix.com/documentation/3.0/manual/discovery/network_discovery/rule

付録A　FAQフォーラム参加特典について

本書の購入者限定特典としてFAQフォーラム（Googleグループ）を用意しています。GoogleアカウントにログインのうえドのURLにアクセスし、本フォーラムの「メンバー登録を申し込む」をクリックしてください。申し込みの際に追加情報として購入した書籍名をご入力ください。

https://groups.google.com/d/forum/vtj-openstack-faq

本フォーラムは書籍の購入者限定のサービスです。フォーラム内のすべての質問に対して日本仮想化技術の社員が回答を行うことをお約束するものではないこと、サービスの公開期間は書籍出版後2年間（〜2018/07）を予定していること、前記期間内であってもOpenStackおよび関連サービスの仕様変更に伴いフォーラムを継続できなくなる可能性があることをご了承ください。また、本書およびフォーラムの解説内容によって生じる直接または間接被害について、著者である日本仮想化技術株式会社ならびに株式会社インプレスでは一切の責任を負いかねます。

●著者紹介

平 愛美
日本仮想化技術株式会社
1983 年生まれ、熊本県出身。崇城大学 (旧：熊本工業大学) 工学部卒業。SIer で Linux サーバーの運用・構築エンジニアなどを経験し、2016 年 4 月より日本仮想化技術株式会社で OpenStack エンジニアに転身。著書に「Linux システム管理標準教科書」や技術誌への寄稿 が多数ある。最近の趣味は自宅 IoT の研究とカメラ。

●スタッフ
- 田中 佑佳（表紙デザイン）
- 鈴木 教之（編集、紙面レイアウト）

本書のご感想をぜひお寄せください
http://book.impress.co.jp/books/1116101031

アンケート回答者の中から、抽選で商品券（1万円分）や図書カード（1,000円分）などを毎月プレゼント。
当選は商品の発送をもって代えさせていただきます。

●本書の内容に関するご質問は、書名・ISBN・お名前・電話番号と、該当するページや具体的な質問内容、お使いの動作環境などを明記のうえ、インプレスカスタマーセンターまでメールまたは封書にてお問い合わせください。電話やFAX等でのご質問には対応しておりません。なお、本書の範囲を超える質問に関しましてはお答えできませんのでご了承ください。

●落丁・乱丁本はお手数ですがインプレスカスタマーセンターまでお送りください。送料弊社負担にてお取り替えさせていただきます。但し、古書店で購入されたものについてはお取り替えできません。

■読者の窓口
インプレスカスタマーセンター
〒101-0051 東京都千代田区神田神保町一丁目105番地
TEL　03-6837-5016　／　FAX　03-6837-5023
info@impress.co.jp

■書店／販売店のご注文窓口
株式会社インプレス 受注センター
TEL　048-449-8040
FAX　048-449-8041

OpenStack構築手順書 Mitaka版（Think IT Books）

2016年7月11日　初版発行

著　者　日本仮想化技術株式会社
発行人　土田　米一
編集人　髙橋　隆志
発行所　株式会社インプレス
　　　　〒101-0051　東京都千代田区神田神保町一丁目105番地
　　　　TEL　03-6837-4635（出版営業統括部）
　　　　ホームページ　http://book.impress.co.jp/

本書は著作権法上の保護を受けています。本書の一部あるいは全部について（ソフトウェア及びプログラムを含む）、株式会社インプレスから文書による許諾を得ずに、いかなる方法においても無断で複写、複製することは禁じられています。

Copyright © 2016 Virtual Tech JP. All rights reserved.
印刷所　京葉流通倉庫株式会社
ISBN978-4-8443-8116-7　C3055
Printed in Japan